T0155555

SpringerBriefs in Electrical and Computer Engineering

Series editors

Woon-Seng Gan, School of Electrical and Electronic Engineering, Nanyang Technological University, Singapore, Singapore

C.-C. Jay Kuo, University of Southern California, Los Angeles, CA, USA

Thomas Fang Zheng, Research Institute of Information Technology, Tsinghua University, Beijing, China

Mauro Barni, Department of Information Engineering and Mathematics, University of Siena, Siena, Italy

SpringerBriefs present concise summaries of cutting-edge research and practical applications across a wide spectrum of fields. Featuring compact volumes of 50 to 125 pages, the series covers a range of content from professional to academic. Typical topics might include: timely report of state-of-the art analytical techniques, a bridge between new research results, as published in journal articles, and a contextual literature review, a snapshot of a hot or emerging topic, an in-depth case study or clinical example and a presentation of core concepts that students must understand in order to make independent contributions.

More information about this series at http://www.springer.com/series/10059

Fen Hou • Yingying Pei • Jingyi Sun

Mobile Crowd Sensing: Incentive Mechanism Design

 Springer

Fen Hou
Department of Electrical and Computer
Engineering
University of Macau
Macau, SAR China

Yingying Pei
Department of Electrical and Computer
Engineering
University of Macau
Macau, SAR China

Jingyi Sun
Department of Electrical and Computer
Engineering
University of Macau
Macau, SAR China

ISSN 2191-8112   ISSN 2191-8120   (electronic)
SpringerBriefs in Electrical and Computer Engineering
ISBN 978-3-030-01023-2   ISBN 978-3-030-01024-9   (eBook)
https://doi.org/10.1007/978-3-030-01024-9

Library of Congress Control Number: 2018956570

This Springer imprint is published by the registered company Springer Nature Switzerland AG
The registered company address is: Gewerbestrasse 11, 6330 Cham, Switzerland

To my parents, my husband, and my kids
— Fen Hou

To my parents, my elder sister, and my boy-friend
— Yingying Pei

To my parents, my litter sister, and my future lover
— Jingyi Sun

Preface

In recent years, we have been witnessing several new trends. Firstly, the advance in hardware technology makes various types of sensors be widely embedded in mobile devices. Secondly, the number of mobile users is explosively growing worldwide. Thirdly, the usage of mobile apps has become increasingly prevalent. Finally, cloud computing platform enables the rapid processing of large-scale data. These trends drive the development of mobile crowd sensing (MCS). Taking the advantages of the sensing function of mobile devices and the mobility of users, MCS is a promising paradigm to efficiently collect data by leveraging a large number of individuals to participate. MCS can be utilized in different areas, such as transportation system, environment monitoring, and traffic control.

In order to achieve the success in MCS, several challenging issues should be addressed. In this book, we focus on the incentive mechanism design to efficiently motivate smartphone users to join in an MCS system such that the system can collect enough number of high-quality sensing data. We start in Chap. 1 by introducing the development and features of MCS. In addition, different applications of MCS and research challenges are elaborated in detail. Chapter 2 provides the basic background knowledge about the auction theory and incentive mechanism design. In Chaps. 3 and 4, we discuss two types of incentive mechanisms: reputation-aware incentive mechanism and social-aware incentive mechanism, respectively. Finally, in Chap. 5, we make a summary and discuss the future work.

This book is intended as a reference for researchers in the field of wireless communications and networking, for electrical and computer engineers who need to design incentive mechanisms and allocate network resources, and for graduate students and senior undergraduate students who are interested in mobile crowd sensing.

We thank the editor, Susan Lagerstrom-Fife, for providing us with the useful links and instructions, which makes the whole process of preparing the manuscript go smoothly. We thank the editor, Caroline Flanagan, for her assistance in handling the publication of this book. FH would like to express her sincere appreciation to Professor Xuemin (Sherman) Shen for his uninterrupted support and help in her career and the preparation of the manuscript. This work is supported by the

University of Macau under grant MYRG2016-00171-FST, the Macau Science and Technology Development Fund under FDCT 121/2014/A3, and joint fund by the Ministry of Science and Technology of the People's Republic of China and the Macau Science and Technology Development Fund under grants 037/2017/AMJ and 020/2014/AMJ.

Macao SAR, China   Fen Hou
Macao SAR, China   Yingying Pei
Macao SAR, China   Jingyi Sun
July 2018

Contents

Chapter 1
Introduction to Mobile Crowd Sensing

1.1 Development of Mobile Crowd Sensing

In recent years, we have been witnessing the explosive growth of mobile users with sensor-embedded smartphones. Firstly, with the development of wireless communication technology and the increase in personal income, mobile devices (e.g., smartphone, ipad, PDA, etc.) have been becoming more and more popular. For instance, smartphones have become the central communication device in people's daily lives [1]. Based on the latest Ericsson mobility report [2], the number of worldwide mobile subscriptions has reached 7.5 billion in 2017 and will approach 9.1 billion in 2022. Secondly, with the development of hardware technology, the size of sensors has become smaller and smaller. Moore's law [3] tells that the number of transistors in a densely integrated circuit doubles about every 18 months. With the miniaturization of sensors, most mobile devices are embedded with various types of sensors (e.g., GPS, accelerometer, camera, digital compass, magnetometer, barometer, gyroscope) [4]. Thus, devices such as smartphones, smart wearable devices (e.g., Google glasses, Apple Watch and Mi Band), unmanned vehicle, and in-vehicle sensing devices (GPS, OBD) can be used to collect information from the surrounding environment. We can use these devices to sense the noise level, traffic situation, temperature, etc. Meanwhile, these devices can upload the sensed data to the server of a data collector through wireless access networks such as cellular system or WiFi at the convenience time and location.

The aforementioned trends motivate the development of the mobile crowd sensing (MCS) [4], which provides an efficient way to collect the sensing data. MCS is a paradigm involving a large number of individuals with sensing and computing devices, in which the individuals can collect the sensing data and extract some specific information. Based on the level of users' involvement, MCS can be classified into two categories: opportunistic sensing [5] and participatory sensing [6]. The former is a passive process, where the sensing data is collected

F. Hou et al., *Mobile Crowd Sensing: Incentive Mechanism Design*,
SpringerBriefs in Electrical and Computer Engineering,
https://doi.org/10.1007/978-3-030-01024-9_1

automatically without the awareness of individuals. The latter is an active process, where individuals need to directly involve in the conducting of sensing activities such as the selection of sensing tasks and the decision of sensing efforts.

Recently, MCS has been attracting more and more attentions in both industry and academia. Several novel service models have been designed with practical implementation. For instance, in the area of mobile market research, Gigwalk [7] and Field Agent [8] are two popular platforms to collect data using MCS. Gigs can be generated and posted on a platform and mobile users can select and conduct some gigs and report the results to the server of the data collector. Similarly, Field Agent is a platform to provide the location-based information such as the collection of in-store information and shopper insights. These information is very useful to help companies to manage and monitor the sale of their products. In addition, Waze [9] is a community-based traffic monitoring platform to provide the real-time report about the traffic condition and road information.

In the literature, a lot of research works have been devoted to different topics in MCS. From the mobile users' point of view, the related topics include the optimal decision of mobile users, the task selection, etc. First of all, how to make the optimal decision about the implementation of sensing tasks is one of the most important issues for mobile users. Cheung et al. in [10] investigate the participation and data reporting of mobile users for the delay-sensitive application, where the MCS system is modeled as a two-stage problem. The service provider decides its optimal reward with the objective of maximizing the achieved profit while the mobile users make their participation and reporting decisions to maximize their own payoffs. Meanwhile, the social relationship of mobile users is widely considered in MCS. Bermejo et al. in [11] investigate the impacts of mobile users' social tie on the performance of data collection in MCS. Cheung et al. in [12] study the impacts of both the social relationship and diversity of mobile users on the collection of high-quality sensing data. The close form about the optimal reward decision of the service provider and the Nash equilibrium of mobile users about the level of their sensing efforts are analyzed. In addition, task selection is another important topic for mobile users in MCS. Cheung et al. in [13] propose a distributed task selection algorithm to collect the time-sensitive and location-dependent sensing data in MCS, where mobile users themselves select sensing tasks according to their own locations, movement costs, movement speeds, and the locations of tasks. Wang et al. in [14] propose a distributed task selection algorithm for mobile users to select their sensing tasks with the objective of maximizing their profits.

From the MCS platform's point of view, one of the most important issues is to motive mobile users to participate in an MCS system such that enough sensing data can be collected. A lot of works have focused on the incentive mechanism design in MCS, which will be elaborated in Chap. 2. Meanwhile, how to keep the long-term participation of mobile users also plays a key role in maintaining enough number of participants. Lee and Hoh in [15] address the participants' drop-off problem and avoid the cost explosion by proposing a dynamic pricing with virtual credit. Gao et al. in [16] propose a Lyapunov based VCG auction method to select mobile users to conduct sensing tasks for maintaining the participation of enough mobile users in a time-dependent and location-aware mobile sensing system. In addition, how to allocate sensing tasks to appropriate mobile users is also a key issue in MCS. Zhao et al. in [17] formulate the task allocation as a min-max aggregate sensing time problem for achieving a fair and energy efficient task allocation. Zhou et al. in [18] propose an efficient task allocation method to maximize the expected sensing revenue by considering the budget constraint and the quality of information for different mobile users. He et al. in [19] propose a near-optimal algorithm to maximize the reward of the MCS platform by jointly considering the locations of tasks and traveling distance constraints of mobile users.

1.2 Features of Mobile Crowd Sensing

Taking the advantages of the smartphones' sensing function and users' mobility, MCS has several unique features such as high flexibility, large scalability, and loose power constraint. Some key features of MCS are elaborated as follows.

- *High scalability*: High scalability is one of the key features for MCS. In a fixed sensor network, enlarging a sensing area will lead to the deployment of sensors at the extended area, which is time-consuming and high cost. However, with MCS, it is easy to achieve the collection of sensing information over a large area by motivating more mobile users at the extended area to participate in the conduction of sensing tasks. Therefore, MCS is an efficient paradigm with high scalability and very suitable to collect sensing data over a large area. Meanwhile, the high scalability also makes MCS more robust compared to a traditional sensor network.
- *Large flexibility*: Another unique feature of MCS is the large flexibility. In a fixed sensor network, the change of sensing area will lead to the adjustment and redeployment of sensors. However, with MCS, what the system needs to do is to recruit some mobile users at the target area to conduct the sensing task, which is easy to achieve due to the availability of a large amount of mobile users. Therefore, MCS is a flexible paradigm to collect sensing data.
- *Self-determination of users' behaviors*: Self-determination of users' behavior is a main feature of the participatory sensing, which is the main category of MCS. In the participatory sensing, mobile users can make their own decisions, rather than being controlled by a central controller. Mobile users are heterogeneous. They may be at different locations with different interests and capabilities. Their mobile devices may have different functions, memory capacity, power, etc. Therefore, each mobile user can decide whether or not to participate in the MCS and take different tasks based on their own situations. Meanwhile, the achieved utility of each mobile user may be impacted by the decisions of other mobile users, which make the self-determination of each mobile user more complex and challenging.

- *Diverse data quality*: Due to the heterogeneity of mobile users, diverse data quality is another feature in MCS. In a fixed sensor networks, sensors are usually pre-deployed based on different objectives. Therefore, the amount of sensors and the type of data they can produce are known as a priori. Thus, it is easier to control the data quality. However, with MCS, the data quality may change with time due to the mobility of mobile users, the preferences of mobile users, and different sensing contexts. Moreover, to preserve participants' privacy, some applications remove the identifying information from the sensor data. Thus, anonymous users may send low quality or even fake data to the platform. In addition, data contributed by different participants may be redundant or inconsistent. Therefore, how to collect high-quality data is a critical issue in MCS due to the diverse data quality.
- *Loose power constraint*: In fixed sensor networks, due to the inconvenience of replacing a sensor's battery, power constraint is a main concern. Many researches are focused on how to prolong the lifetime of the network. However, with MCS, sensors are embedded in users' mobile devices. Users can be aware of the remaining battery anytime and charge their devices conveniently when the battery is below a certain level. Therefore, power constraint is not the main concern in MCS.

1.3 Applications of Mobile Crowd Sensing

With a low deployment cost and several nice features, MCS has been considered and utilized in many areas such as transportation, environment monitoring, healthy care and well-being, social recommendation applications, etc. Meanwhile, many corresponding systems and applications have been designed and developed in these areas. We introduce some existing and potential MCS applications as follows.

1.3.1 Transportation

Crowd sensing data can be used both for public infrastructure measuring or individual travel planning. Public infrastructure measuring applications include the measuring of traffic congestion, road condition or parking availability [20]. For example, Mohan et al. in [21] design Nericell to monitor road and traffic conditions. Nericell uses various sensors embedded in smartphones (e.g., GSM radio, GPS, accelerometer, etc.) to detect the specific road and traffic conditions such as bumps, potholes, honking and braking, etc. Thiagarajan et al. in [22] propose Vtrack, which uses a variety of sensors (e.g., GPS, WiFi, and/or cellular triangulation) to measure and localize traffic delay and congestion, then provide the efficient individual travel route for users. For example, it is used to suggest personal driving routes or decide the departure time. Mobile Millennium [23] is a research project that uses GPS in mobile phones to gather a large scale traffic information, process it, and return fine-grained traffic information back to mobile users in real time. GreenGPS [24] uses MCS data to map fuel consumption on city streets, allowing drivers to find the most fuel-efficient routes for their vehicles between arbitrary end-points.

1.3.2 Environment Monitoring

MCS provides a new way for environment monitoring such as the measuring of the air pollution, the estimation of the noise level, the preservation of the nature, etc.

- *Air pollution*: Air pollution is a worldwide issue, the World Health Organization reports that seven million premature deaths annually are linked to air pollution [25]. The PEIR project [26] uses location data sampled from smartphones to calculate the personalized estimate of environmental impact and exposure. Common Sense [27] allows individuals to measure their personal air pollutants, groups to aggregate their members' exposure, and activists to mobilize grassroots community action. Kanjo et al. in [28] design MobGeoSen, which enables individuals to monitor their local environment (e.g., pollution and temperature) and their private spaces (e.g., activities and health) by using smartphones in their daily lives.
- *Noise pollution*: Maisonneuve et al. in [29] propose a new method named NoiseTube to measure and map the noise pollution, which encourages citizens with GPS and noise sensor equipped in smartphones to measure their personal collected noise information in their daily environment. With the shared geo-localized measurements from different smartphone users, a collective noise map can be built up. Rana et al. in [30] design an end-to-end participatory urban noise mapping system named EarPhone to recover the noise map based on the sensing data collected by the crowd. It is used to detect the environmental noise level, especially the roadside ambient noise.

1.3.3 Health and Well-Being

Health-related applications mainly include health monitoring and health manage-ment. By collecting and analyzing the real-time biomedical and environmental data of participants, it is very helpful for the participants to find the possible reasons of some chronic illness. SPA [31] is such a smartphone assisted chronic illness self-management crowd sensing system. In addition, Denning et al. in [32] have designed BALANCE to detect users' caloric expenditure automatically. The sensor data is collected by a smartphone sensing platform, which is a device worn on the hip of participants. Hicks et al. in [33] have designed AndWellness to collect and analyze the personal data from both active, triggered user experience samples and passive logging of onboard environmental sensors. The analysis result can be observed by both researchers and the participants in real-time. Besides the aforementioned health monitoring services, MCS data is also used for personal well-being management. For example, DietSense system [34] allows people to photograph and share their plates during the mealtime. By comparing the eating habits of people within a community, they can control their diet and provide suggestions to others. De Oliveira and Oliver in [35] propose an MCS system TripleBeat to encourage runners to achieve predefined exercise goal by musical feedback.

1.3.4   Mobile Social Recommendation

There are two main research trends in mobile social recommendation: friend recommendation and social activity recommendation [36]. By exploring MCS data, a number of recommendation systems are enabled to provide personalized services. By calculating the similarity among users, it is reasonable to recommend one's preference to another similar user. For example, GeoLife [37] provides an individual with locations matching her travel preferences by mining multiple users' GPS traces. Social activity recommendation applications provide group activity organization and recommendation. For instance, MobiGroup [38] is a group-aware MCS system by leveraging features extracted from both online and offline communities.

1.4   Research Challenges

The main research challenges in MCS include incentive mechanism design, the quality and redundancy of data, privacy and security, etc.

- *Incentive mechanism design*: The objective of MCS is to collect enough high-quality sensing data. However, the conduction of sensing tasks makes either implicit efforts (e.g., monetary costs) or explicit efforts (e.g. give some input or assessments) [36]. Meanwhile, the sensor-equipped mobile devices are owned and controlled by individual users. Therefore, they may not be willing to contribute their sensing capabilities unless they can receive appropriate compensation for their consumption caused by the conduction of sensing tasks. In order to achieve the goal of collecting enough sensing data, an efficient incentive mechanism is essential for the success of MCS. The incentives are usually divided into three categories: interest and entertainment, social and ethical, financial rewards. The financial rewards could be money, virtual cash, or credits. However, once money is involved, the participants are more likely to manipulate the system by providing some fake information. Therefore, truthful incentive mechanisms should be designed to ensure the quality of the collected sensing data and avoid that some mobile users manipulate the MCS system.

- *Activity optimization for mobile users*: Given enough number of participants in MCS, their activities such as the selection of sensing tasks and the reporting of sensing data are also important to both users and the system. Considering the heterogeneity of mobile users, on one hand, the heterogeneous mobile users compete with each other; on the other hand, they can cooperate with each other to reduce the cost or complete some complex tasks. How to model and analyze the competition and cooperation of mobile users is one of the main challenging issues in achieving the group or individual activity optimization of mobile users. In addition, the data redundancy and data diversity should be considered in optimizing the mobile users' behavior. For instance, nearby mobile users may send similar data in the collection of noise level, which leads to data redundancy. In addition, devices at the same location may have different sensing capacities due to the difference of mobile brand or the different configuration. Even the similar devices may get different sensing data when they are under different conditions (e.g., the device in a pocket or out of the pocket). Therefore, from the system point of view, how to optimize the mobile users' behaviors is also an important issue.
- *Privacy and security*: In MCS applications, the concern about the privacy or security may discourage users from data sharing. For example, the disclosure of users' identity or some sensitive attributes such as locations (e.g. home address or work locations), daily commutes routs and personal activities or conditions. On the other hand, some MCS tasks cannot be executed without the sensitive sensor data. For example, in noise pollution measuring application, it is necessary to read the GPS sensor and microphone sensor at the same time. Therefore, exploring approaches that can both ensure the participants' quantity and also preserves the privacy of participants is necessary and challenging. A popular technique is anonymization [39], which removes identifying information from sensor data before revealing it to a third party. However, the drawback is that the anonymous user may provide incorrect or even fake data to the system. In this case, we have to develop privacy protected and quality ensured technologies to collect data in MCS.

References

1. Lane ND, Miluzzo E, Lu H, Peebles D, Choudhury T, Campbell AT (2010) A survey of mobile phone sensing. IEEE Communications magazine 48(9)
2. Heuveldop N (2017) Ericsson mobility report. Ericsson, Stockholm
3. Schaller RR (1997) Moore's law: past, present and future. IEEE spectrum 34(6):52–59
4. Ganti RK, Ye F, Lei H (2011) Mobile crowdsensing: current state and future challenges. IEEE Communications Magazine 49(11)
5. Khan WZ, Xiang Y, Aalsalem MY, Arshad Q (2013) Mobile phone sensing systems: A survey. IEEE Communications Surveys & Tutorials 15(1):402–427

6. Burke JA, Estrin D, Hansen M, Parker A, Ramanathan N, Reddy S, Srivastava MB (2006) Participatory sensing. In Proc. of ACM Conference on Networked Sensor Systems, Boulder, Colorado, USA, October 2006.

7. Gigwalk, "https://urldefense.proofpoint.com/"

8. Fieldagent, "https://urldefense.proofpoint.com"

9. Waze, "https://urldefense.proofpoint.com"

10. Cheung MH, Hou F, Huang J (2018) Delay-Sensitive Mobile Crowdsensing: Algorithm Design and Economics. IEEE Transactions on Mobile Computing

11. Bermejo C, Chatzopoulos D, Hui P (2016) How sustainable is social based mobile crowd-sensing? An experimental study. In: Network Protocols (ICNP), 2016 IEEE 24th International Conference on, 2016. IEEE, pp 1–6

12. Cheung MH, Hou F, Huang J (2017) Make a difference: Diversity-driven social mobile crowdsensing. In: INFOCOM 2017-IEEE Conference on Computer Communications, IEEE, 2017. IEEE, pp 1–9

13. Cheung MH, Southwell R, Hou F, Huang J (2015) Distributed time-sensitive task selection in mobile crowdsensing. In: Proceedings of the 16th ACM International Symposium on Mobile Ad Hoc Networking and Computing, 2015. ACM, pp 157–166

14. Z. Wang, J. Hu, J. Zhao, D. Yang, H. Chen, and Q. Wang (2018) Pay on-demand: Dynamic incentive and task selection for location-dependent mobile crowdsensing systems, IEEE International Conference on Distributed Computing Systems, Vienna, Austria, July 2018.

15. Lee J-S, Hoh B (2010) Sell your experiences: a market mechanism based incentive for participatory sensing. In: Pervasive Computing and Communications (PerCom), 2010 IEEE International Conference on, 2010. IEEE, pp 60–68

16. Gao L, Hou F, Huang J (2015) Providing long-term participation incentive in participatory sensing. In: Computer Communications (INFOCOM), 2015 IEEE Conference on, 2015. IEEE, pp 2803–2811

17. Zhao Q, Zhu Y, Zhu H, Cao J, Xue G, Li B (2014) Fair energy-efficient sensing task allocation in participatory sensing with smartphones. In: INFOCOM, 2014 Proceedings IEEE, 2014. IEEE, pp 1366–1374

18. Zhou C, Tham C-K, Motani M (2015) QOATA: QoI-aware task allocation scheme for mobile crowdsensing under limited budget. In: Intelligent Sensors, Sensor Networks and Information Processing (ISSNIP), 2015 IEEE Tenth International Conference on, 2015. IEEE, pp 1–6

19. He S, Shin D-H, Zhang J, Chen J (2017) Near-optimal allocation algorithms for location-dependent tasks in crowdsensing. IEEE Transactions on Vehicular Technology 66(4):3392–3405

20. Mathur S, Jin T, Kasturirangan N, Chandrasekaran J, Xue W, Gruteser M, Trappe W (2010) Parknet: drive-by sensing of road-side parking statistics. In: Proceedings of the 8th international conference on Mobile systems, applications, and services, 2010. ACM, pp 123–136

21. Mohan P, Padmanabhan VN, Ramjee R (2008) Nericell: rich monitoring of road and traffic conditions using mobile smartphones. In: Proceedings of the 6th ACM conference on Embedded network sensor systems, 2008. ACM, pp 323–336

22. Thiagarajan A, Ravindranath L, LaCurts K, Madden S, Balakrishnan H, Toledo S, Eriksson J (2009) VTrack: accurate, energy-aware road traffic delay estimation using mobile phones. In: Proceedings of the 7th ACM conference on embedded networked sensor systems, 2009. ACM, pp 85–98

23. UC Berkeley/Nokia/NAVTEQ, Mobile Millennium. "http://traffic.berkeley.edu/"

24. Ganti RK, Pham N, Ahmadi H, Nangia S, Abdelzaher TF (2010) GreenGPS: a participatory sensing fuel-efficient maps application. In: Proceedings of the 8th international conference on Mobile systems, applications, and services, 2010. ACM, pp 151–164

25. World Health Organization. "http://www.who.int/"

26. Mun M, Reddy S, Shilton K, Yau N, Burke J, Estrin D, Hansen M, Howard E, West R, Boda P (2009) PEIR, the personal environmental impact report, as a platform for participatory sensing systems research. In: Proceedings of the 7th international conference on Mobile systems, applications, and services, 2009. ACM, pp 55–6

27. Dutta P, Aoki PM, Kumar N, Mainwaring A, Myers C, Willett W, Woodruff (2009) A Common sense: participatory urban sensing using a network of handheld air quality monitors. In: Proceedings of the 7th ACM conference on embedded networked sensor systems, 2009. ACM, pp 349–350

28. Kanjo E, Benford S, Paxton M, Chamberlain A, Fraser DS, Woodgate D, Crellin D, Woolard A (2008) MobGeoSen: facilitating personal geosensor data collection and visualization using mobile phones. Personal and Ubiquitous Computing 12(8):599–607

29. Maisonneuve N, Stevens M, Niessen ME, Steels L (2009) NoiseTube: Measuring and mapping noise pollution with mobile phones. In: Information technologies in environmental engineering. Springer, pp 215–228

30. Rana RK, Chou CT, Kanhere SS, Bulusu N, Hu W (2010) Ear-phone: an end-to-end participatory urban noise mapping system. In: Proceedings of the 9th ACM/IEEE International Conference on Information Processing in Sensor Networks, 2010. ACM, pp 105–116

31. Sha K, Zhan G, Shi W, Lumley M, Wiholm C; Arnetz B (2008) SPA: a smart phone assisted chronic illness self-management system with participatory sensing. In: Proceedings of the 2nd International Workshop on Systems and Networking Support for Health Care and Assisted Living Environments, 2008. ACM, p 5

32. Denning T, Andrew A, Chaudhri R, Hartung C, Lester J, Borriello G, Duncan G (2009) BALANCE: towards a usable pervasive wellness application with accurate activity inference. In: Proceedings of the 10th workshop on Mobile Computing Systems and Applications, 2009. ACM, p 5

33. Hicks J, Ramanathan N, Kim D, Monibi M, Selsky J, Hansen M, Estrin D (2010) AndWellness: an open mobile system for activity and experience sampling. In: Wireless Health 2010, 2010. ACM, pp 34–43

34. Reddy S, Parker A, Hyman J, Burke J, Estrin D, Hansen M (2007) Image browsing, processing, and clustering for participatory sensing: lessons from a DietSense prototype. In: Proceedings of the 4th workshop on Embedded networked sensors, 2007. ACM, pp 13–17

35. De Oliveira R, Oliver N (2008) TripleBeat: enhancing exercise performance with persuasion. In: Proceedings of the 10th international conference on Human computer interaction with mobile devices and services, 2008. ACM, pp 255–264

36. Guo B, Wang Z, Yu Z, Wang Y, Yen NY, Huang R, Zhou X (2015) Mobile crowd sensing and computing: The review of an emerging human-powered sensing paradigm. ACM Computing Surveys (CSUR) 48(1):7

37. Zheng Y, Xie X, Ma W-Y (2010) Geolife: A collaborative social networking service among user, location and trajectory. IEEE Data Eng Bull 33(2):32–39

38. Guo B, Yu Z, Chen L, Zhou X, Ma X (2016) MobiGroup: Enabling lifecycle support to social activity organization and suggestion with mobile crowd sensing. IEEE Transactions on Human-Machine Systems 46(3):390–402

39. Sweeney L (2002) k-anonymity: A model for protecting privacy. International Journal of Uncertainty, Fuzziness and Knowledge-Based Systems 10(05):557–570

Chapter 2
Auction Theory and Incentive Mechanism Design

2.1 Auction Theory

Auction theory is a branch of economics which deals with how people sell or buy objects. The history of auction can be date back to 500 BC, at which time it was recorded that auctions of women for marriage were held annually in Babylon [1]. Nowadays, numerous kinds of commodities such as antiques, arts, bonds, wireless spectrum, and so on, are sold by means of auction. For instance, auctions are used to sale and disposal of assets such as the open bidding for government procurement, construction works, and right to natural resources (e.g., oil, natural gas, etc.). In recent years, auction has been widely used in spectrum resource allocation, e-commerce websites, online advertising, etc.

An auction system usually includes three parts: buyer, seller, and auctioneer. In general, there are four basic types of auctions: First-price sealed bid auction, second-price sealed bid auction, English auction, and Dutch auction, where the English auction and Dutch auction belong to the category of the open-bid auction, and the other two auctions belong to the category of sealed-bid auction [1].

English auction is also called open-cry ascending-bid auction, where the auctioneer calls an initial low price and then gradually raises the price until there remains only one buyer. Then, this buyer wins the objects and pays the current bid price. This type of auction is often used to sell artwork, land, etc.

Dutch auction is also called open-cry descending bid auction, where the auctioneer calls an initial high price and gradually lowers the price until one buy accepts this current price. Then, this buyer wins the object and pays the current price. This type of auction is used to figure out the optimum price for a stock in an initial public offering. It is also suitable for selling perishable communities such as vegetables and fruits.

© The Author(s), under exclusive license to Springer Nature Switzerland AG 2019
F. Hou et al., *Mobile Crowd Sensing: Incentive Mechanism Design*,
SpringerBriefs in Electrical and Computer Engineering,
https://doi.org/10.1007/978-3-030-01024-9_2

In a first-price sealed-bid auction, buyers simultaneously hand their sealed bid prices to the auctioneer. Then, the buyer with the highest bid price wins the auction and pays exact amount that the winner bids.

In a second-price sealed-bid auction, buyers also simultaneously hand their sealed bid prices to the auctioneer. Then, the buyer with the highest bid price wins the auction, but the winner pays the second highest bid price.

To avoid an object for being sold at an unexpectedly low price, a seller may set a reserve price for these four types of auctions. In addition, in many situations, each buyer knows the value of the object to himself while no buyer knows the other bidders' evaluation on this object. In that case, the Dutch open descending price auction is strategically equivalent to the first-price sealed-bid auction, and the English open ascending auction is also equivalent to the second-price sealed bid auction [1].

2.2 Incentive Mechanism Design

Mechanism design belongs to the field of reverse engineering economics. It deals with problems in which multiple rational agents are to be organized in a way that the global outcome meets desired objectives and properties such as truthful, Pareto optimality, individual rationality, and computational efficiency. Mechanism design generally provides rules for the coordination of agents in distributed decision problems. For a system including a central controller, it is feasible for the central controller to arrange and control the behavior of each component of this system. However, for a system without a central controller, each component of this system can make its own decision. In this case, a properly designed mechanism seems as replacing a central controller: although each component makes its own decision, the whole system can work properly and achieve the objectives of the designer. So, the mechanism design plays a key role in making a system work efficiently without a control of a central part. The main objective of MCS is to collect enough number of sensing data. Therefore, how to incentive mobile users to join in the MCS is one of the most important and challenging issues. Meanwhile, MCS is usually a system or platform without a central controller. Mobile users in an MCS system can make their own decisions based on their preferences and situations. Therefore, incentive mechanism design plays an essential role in the success of MCS such that both the mobile users and the data collector in an MCS system can achieve their own objectives. In specific, the data collector can obtain enough number of high-quality sensing data with low cost, while mobile users can achieve the desired reward to compensate their consumption in the conduction of sensing tasks.

Besides MCS, incentive mechanism design is also a key research issue in other types of networks such as ad-hoc network, peer-to-peer (P2P) network, etc. For instance, the cooperative communication is one way to improve the capacity of a wireless network. The main obstacle of cooperative communication is the lack of relay nodes. So incentive mechanism can be designed for cooperative communication to incentive more participating nodes to work as relay nodes. Yang et al. in [2] propose a truthful auction scheme for cooperative communications to stimulate wireless nodes working as a relay so that the capacity of a wireless network can be improved. In this proposed scheme, the relay nodes work as sellers who sell their relay services with associated price. The source nodes work as buyers who purchase these relay services. And the base station works as the auctioneer who determines the winners and the payment.

The incentive mechanism is also designed for peer-to-peer network. Habib and Chuang in [3] propose an incentive mechanism for a peer-to-peer media streaming system. The proposed mechanism is ranked based and can provide the service differentiation in peer selection for P2P streaming. That is, the system will reward the good contributors with more choice or flexibility in peer selection process such that the contributors will get better streaming sessions with higher quality. But free-riders will have few opportunities to select the peer. So they will get low quality streaming.

Mahmoud and Shen in [4] design an incentive mechanism for multi-hop cellular network, which combines the characteristic of ad hoc network with cellular system to improve the performance of the cellular network. Because the multi-hop cellular network involves many autonomous devices in the process of forwarding packet and routing, there will be the security problem, the selfish device don't want to relay other nodes' packets because cooperation will consume their own resource and cannot provide any immediate advantage for them. This situation will degrade the efficiency and throughput of the network. The proposed incentive mechanism is to stimulate the selfish nodes to cooperate with each other. Based on the proposed mechanism, the forwarding nodes will get the reward. The proposed mechanism is robust and can improve the performance of the multi-hop cellular network very well.

2.3 Applications of Auction in Incentive Mechanism Design

Auction theory is a subset of mechanism design. An MCS system consists of three parts: the service provider (SP), smartphone users (SUs), and the platform. From the ecosystem's point of view, an MCS system can be modeled as an auction system, where the sensing data are considered as the goods, SUs are the sellers who produce and sell the goods. The SP is the buyer who purchases the sensing data from the sellers. The sellers (SUs) sell their goods (sensing data) to buyers (SPs) and will get payment. In this system, the number of SP could be one or more. When an SP wants to collect the sensing data, she can post the sensing task on the platform, appended with the task description and other information. The platform could be some mobile phone apps, which could be developed and controlled by a third-party. The sensing tasks on the platform are only visible for the SUs who download and install these apps. An SU can apply to conduct these sensing tasks. If the platform accepts the application, she will send a message to the SU. Then the SU will conduct the sensing task and upload the sensing data. After collecting enough sensing data, the SP can conduct the data analysis, obtain a certain convictive result for some phenomena, and provide corresponding applications and services. However, as potential participants, SUs can make their own decisions about participating in the sensing process or not. Therefore, how to use auction theory to design an efficient incentive mechanism to motivate more SUs to conduct the sensing task plays a critical role in the success of an MCS system.

In an MCS system, the sellers (SUs) report the sensing data and obtain corresponding payment. This payment can be considered as the incentive, which can be a monetary or tangible or virtual reward. The rational SUs always want a larger payment so that they will get a larger utility. So it is crucial to consider the truthfulness of the buyers while revealing the values on a target object in an incentive mechanism. Four main properties are desired for an efficient incentive mechanism, which are given as follows.

• Truthfulness: Each player does not have the incentive to deviate from submitting its own true valuation;
• Individual Rationality: the utility of each buyer is non-negative;
• Budget Balance: the utility of the platform is non-negative;
• Computational Efficiency: the computation complexity is reasonable.

Incentive mechanisms can be specifically designed for achieving various objectives. For economical organizations which pursue interests and benefits, the mechanism designer may only wish to maximize the social welfare. While for a government or other public organizations, their purposes are not only to maximize the social welfare, but also to share the surplus among the users as evenly as possible. Most of the existing work about the incentive mechanism design is focused on maximizing social welfare, which can be measured by $\sum_{i=1}^{N} v_i$, where v_i is the utility of user i and N is the total number of users. The objective of some research work is to maximize the quality of data collected from mobile users. Jaimes et al. in [5] take the location information of SUs into consideration and propose a greedy incentive algorithm to improve the coverage of the area of interest as well as reducing the collection of redundant data. Zhou et al. in [6] design an approximately truthful incentive mechanism to maximize the value of services as well as protecting the users' privacy.

Some works consider the fairness in the incentive mechanism design [7–12], which may sacrifice some utilitarian efficiency. Cole et al. in [9] propose a partial allocation mechanism to incentive the agents to truthfully report their valuations, which can provide every agent with at least a $\frac{1}{e}$ fraction of her proportionally fair valuation. Pai and Vohra in [8] study the government procurement and allocation problem and compare two methods: use subsidies or set-asides for different goals. Wang et al. in [12] propose a generalized AGV mechanism to balance the agents' payoffs in a max-min fair manner, while maintaining the properties of a classical AGV mechanism such as Bayesian incentive compatibility, ex-post efficiency, and ex-post budget balance. Porter et al. in [10] consider the k-fairness. If no agents loses more than $v_{[k]}/n$, a mechanism is said to be k-fairness, where $v_{[k]}$ is the kth lowest cost among all agents.

Incentive mechanism can be designed based on different objectives, such as improving the users' utility, collecting specific location's data, improving the data quality, maximizing the social welfare, etc.

We can address the incentive mechanism design from different aspects. First of all, truthfulness is an important aspect in incentive mechanism design so as to prevent any SU from rigging its bid to manipulate the market [2, 13–18]. In the aspect of design method, the reverse auction is used in [5, 13, 16, 19] to design the incentive mechanism, while double auction theory is used in [15, 17, 18, 20]. The detailed elaboration will be presented as follows.

2.3.1   Reverse Auction Based Mechanisms

Reverse auction is a type of auction in which the roles of buyer and seller are reversed. Usually in the reverse auction, multiple sellers sell goods to a single buyer and the price will typically decrease as the sellers underbid each other. Reverse auction is widely used in the resource allocation. Feng et al. in [16] propose an incentive mechanism named TRAC (Truthful Auction for Location-Aware Collaborative Sensing) for MCS by taking the location information of SUs into consideration when assigning sensing tasks to SUs. A reverse auction method is used to model the interaction between the platform and SUs. TRAC consists of two parts. The first part is to determine the winning bids. In this part, a near-optimal approximation algorithm is designed with low computation complexity. The second part is to determine the payment, which is critical to guarantee the truthfulness. TRAC is truthfulness, individual rationality and computation efficiency.

Lee and Hoh in [19] propose a reverse auction based dynamic price incentive mechanism named RADP, where smartphone users can sell their sensing data to the SP with their claimed bid prices, and then the SP selects multiple users and purchases their sensing data. The proposed incentive mechanism aims to minimize the incentive cost, maintain the adequate number of participants, and prevent users from dropping out of the sensing procedure by giving the losers some virtual credits. But this mechanism can not achieve the truthfulness, which is a very important property since this property ensures users to claim their true cost as the bid price to avoid market manipulation [1].

Yang et al. in [13] propose two incentive mechanisms for the platform-centric model and user-centric model respectively. For the platform-centric model, only one sensing task is considered. The platform announces a total reward which will be shared among the selected SUs, and the SUs can determine their own sensing plan accordingly. Stackelberg game is used to design an incentive mechanism named LSB auction, in which the utility of the platform is maximized by calculating the unique Stackelberg Equilibrium, and no user can improve her utility by unilaterally deviating from her equilibrium strategy. But in this case the LSB auction mechanism is not truthful. For user-centric model, multiple tasks are announced by the platform and each SU can select a subset of tasks. An auction-based incentive mechanism named MSensing is proposed for user-centric model. In MSensing, SUs can select several tasks and submit their claimed bid price to the platform, then the platform will determine the winning SUs. This mechanism can guarantee the truthfulness, individual rationality, profitability and computational efficiency.

Jaimes et al. in [5] present a greedy incentive algorithm by taking the location information of SUs into consideration. The reverse auction is used to design the incentive mechanism. To improve the coverage area, as well as reducing the collection of redundant data, location is considered under the budget and coverage constraints. The greedy algorithm selects a representative subset of SUs according to their locations given a fixed budget. The proposed mechanism is efficient but not truthful.

2.3.2 Double Auction Based Mechanisms

Double auction is another common method used in the incentive mechanism design, where potential buyers and sellers simultaneously submit their bid prices and ask prices, then the auctioneer will choose an appropriate price p to clear the market. Double auction is also widely used in the mechanism design for wireless networks. Sun et al. in [20] propose a social-aware incentive mechanism (SAIM), which is a truthful double auction mechanism to achieve the efficient allocation of sensing tasks in MCS. SAIM considers the social relationship of SUs and satisfies the properties of individual rationality, budget balance, the completely truthful for SPs, and partially truthful for SUs.

Chen and Wang in [15] propose a truthful incentive mechanism named SPARC, which is for multiple sensing tasks and multiple SUs in a participatory sensing system. Each task has a demand of sensing time, and each SU has his own available sensing time. SPARC allocates winning tasks' sensing time among winning users to maximize the social welfare, which can guarantee the truthfulness and improve the sensing time utilization, as well as the participatory sensing task satisfaction ratio.

Wei et al. in [17] propose a truthful online double auction where both SUs and SPs are dynamic. A general framework for designing a truthful online double auction for dynamic MCS is presented and the proposed incentive mechanism can ensure truthfulness, individual rationality and budget balance.

Zhang et al. in [18] study three system models for crowdsourcing and design an incentive mechanism for each model, which are SS-Model, SM-Model, and MM-Model. In SS-model, there is only one requester, and each provider can claim only one bid. In SM-model, there is one requester, and each provider can claim multiple bids but at most one of the sets of tasks she claims will be allocated to her. In MM-model, there are multiple requesters and multiple providers. The requesters work as buyers and the providers work as sellers. Each provider also can claim multiple bids. The authors propose an incentive mechanism for each of them. These incentive mechanisms can achieve individual rationality, budget balance and truthfulness.

2.3.3 Other Truthful Mechanisms

Koutsopoulos in [21] has designed an incentive mechanism for MCS by using a reverse auction. The SP accepts service queries and organizes a reverse auction. Different from other works, users in this system can participate in different capacities. For example, users can submit different types of sensing data or different number of sensing data. The author defines the participation level as the efforts of each user in conducting sensing tasks. Each user reports her unit cost, and the SP determines the participation level and payment of users. This work aims at minimizing the cost of compensating participants and ensure the sensing quality of service requesters. An optimal reverse auction problem is formulated. The proposed mechanism is individually rational and incentive compatible.

Ueyama et al. in [22] propose a novel incentive mechanism based on gamification for MCS. Considering the situation when sensing task is heavy for users, in this case, the stronger incentive is required and the total rewards paid by the client will quickly rise. So this work aims to design an incentive mechanism to reduce the total amount of rewards paid by the client. Different from other works, the proposed incentive mechanism combines several schemes. The first one is a status level scheme. Users with higher status can earn more reward points than the users with lower levels, even if they have completed the same sensing task. Users can reach an upper level by completing more sensing tasks. The second one is a ranking scheme. The system maintains a ranking list of all participants based on their amount of reward points. All participants can access to this rank through the Internet, so users have the motivation to participate in more tasks so that their position in the ranking will be much higher. The third one is a badge scheme. When a user has completed several sensing tasks and satisfied a certain condition, she can receive a badge, which looks more like a medal for this user. The badge scheme can be used to attract more users to participate by the sense of accomplishment rather than the monetary reward. So the proposed incentive mechanism uses both monetary incentive and gamification to incent users to participate in sensing tasks. A heuristic algorithm is designed to select the set of users to conduct the sensing tasks and determine the reward points.

Yang et al. in [2] propose a truthful auction scheme for cooperative communications (TASC) to stimulate wireless nodes to work as relay nodes so that the capacity of the wireless networks can be improved. In this scheme, the relay nodes work as sellers who sell their relay services with associate price. The source nodes work as buyers who purchase these relay services. And the base station works as the auctioneer who determines the winners and the payment.

References

1. Krishna V (2009) Auction Theory, Academic press.
2. Yang D, Fang X, Xue G (2011) Truthful auction for cooperative communications. In: Proceedings of the Twelfth ACM International Symposium on Mobile Ad Hoc Networking and Computing, 2011. ACM, p 9

3. Habib A, Chuang J (2004) Incentive mechanism for peer-to-peer media streaming. In: Quality of Service, 2004. IWQOS 2004. Twelfth IEEE International Workshop on, 2004. IEEE, pp 171–180
4. Mahmoud ME, Shen X (2009) DSC: Cooperation incentive mechanism for multi-hop cellular networks. In: Communications, 2009. ICC'09. IEEE International Conference on, 2009. IEEE, pp 1–6
5. Jaimes LG, Vergara-Laurens I, Labrador MA (2012) A location-based incentive mechanism for participatory sensing systems with budget constraints. In: Pervasive Computing and Communications (PerCom), 2012 IEEE International Conference on, 2012. IEEE, pp 103–108
6. Zhou Y, Zhang Y, Zhong S (2017) Incentive Mechanism Design in Mobile Crowd Sensing Systems with Budget Restriction and Capacity Limit. In: Computer Communication and Networks (ICCCN), 2017 26th International Conference on, 2017. IEEE, pp 1–9
7. d'Aspremont C, Cremer J, Gérard-Varet L-A (1990) Incentives and the existence of Pareto-optimal revelation mechanisms. Journal of Economic Theory 51(2):233–254
8. Pai MM, Vohra R (2012) Auction design with fairness concerns: Subsidies vs. set-asides. Discussion Paper, Center for Mathematical Studies in Economics and Management Science
9. Cole R, Gkatzelis V, Goel G (2013) Mechanism design for fair division: allocating divisible items without payments. In: Proceedings of the fourteenth ACM conference on Electronic commerce, 2013. ACM, pp 251–268
10. Porter R, Shoham Y, Tennenholtz M (2004) Fair imposition. Journal of Economic Theory 118(2):209–228
11. Atlamaz M, Yengin D (2008) Fair groves mechanisms. Social Choice and Welfare 31(4):573–587
12. Wang T, Xu Y, Ahipasaoglu SD, Courcoubetis C (2015) Max-min fairness of generalized AGV mechanisms. In: Decision and Control (CDC), 2015 IEEE 54th Annual Conference on, 2015. IEEE, pp 5170–5177
13. Yang D, Xue G, Fang X, Tang J (2012) Crowdsourcing to smartphones: incentive mechanism design for mobile phone sensing. In: Proceedings of the 18th annual international conference on Mobile computing and networking, 2012. ACM, pp 173–184
14. Xu W, Huang H, Sun Y-e, Li F, Zhu Y (2013) DATA: A double auction based task assignment mechanism in crowdsourcing systems. In: 2013 8th International Conference on Communications and Networking in China (CHINACOM), 2013. IEEE, pp 172–177
15. Chen C, Wang Y (2013) Sparc: Strategy-proof double auction for mobile participatory sensing. In: Cloud Computing and Big Data (CloudCom-Asia), 2013 International Conference on, 2013. IEEE, pp 133–140
16. Feng Z, Zhu Y, Zhang Q, Ni LM, Vasilakos AV (2014) TRAC: Truthful auction for location-aware collaborative sensing in mobile crowdsourcing. In: INFOCOM, 2014 Proceedings IEEE, 2014. IEEE, pp 1231–1239
17. Wei Y, Zhu Y, Zhu H, Zhang Q, Xue G (2015) Truthful online double auctions for dynamic mobile crowdsourcing. In: Computer Communications (INFOCOM), 2015 IEEE Conference on, 2015. IEEE, pp 2074–2082
18. Zhang X, Xue G, Yu R, Yang D, Tang J (2015) Truthful incentive mechanisms for crowdsourcing. In: Computer Communications (INFOCOM), 2015 IEEE Conference on, 2015. IEEE, pp 2830–2838
19. Lee J-S, Hoh B (2010) Sell your experiences: a market mechanism based incentive for participatory sensing. In: Pervasive Computing and Communications (PerCom), 2010 IEEE International Conference on, 2010. IEEE, pp 60–68
20. Sun J, Hou F, Ma S, Shan H (2016) Social-aware incentive mechanism for participatory sensing. In: Global Communications Conference (GLOBECOM), 2016 IEEE, 2016. IEEE, pp 1–6
21. Koutsopoulos I (2013) Optimal incentive-driven design of participatory sensing systems. In: Infocom, 2013 proceedings ieee, 2013. IEEE, pp 1402–1410
22. Ueyama Y, Tamai M, Arakawa Y, Yasumoto K (2014) Gamification-based incentive mechanism for participatory sensing. In: Pervasive Computing and Communications Workshops (PERCOM Workshops), 2014 IEEE International Conference on, 2014. IEEE, pp 98–103

Chapter 3
Reputation-Aware Incentive Mechanism Design

3.1 An Important Factor: Reputation

During the collection of sensing data, the service provider (SP) aims to obtain a large amount of data. However, in addition to the quantity of the collected data, the quality of each collected data is also important. Smartphone users (SUs) may submit some erroneous or unreliable data intentionally or unintentionally, and the low-quality data will impact the accuracy of data analysis result, degrade the satisfaction of an SP. Therefore, the quality of sensing data is one of the most critical factors in the collection of sensing data, and should be considered in the incentive mechanism design.

Reputation is one way to evaluate the quality of data provided by a smartphone user. The reputation of each smartphone user can be updated using a reputation system like the one proposed in [1].

Kantarci and Mouftah in [2] consider the reputation in the collection of sensing data. However, the proposed method in [2] mainly focus on the improvement of the SP's utility. It cannot guarantee the truthfulness. Therefore, we aim to design a truthful incentive mechanism by jointly considering the quality of sensing data. We use the reputation to evaluate the quality of sensing data that a smartphone user submits. It reflects the accuracy of recently reported data. With the consideration of the reputation of smartphone users, we aim to design a truthful incentive mechanism which can maximize the weighted social welfare, and improve the overall quality of the collected data as well.

© The Author(s), under exclusive license to Springer Nature Switzerland AG 2019
F. Hou et al., *Mobile Crowd Sensing: Incentive Mechanism Design*,
SpringerBriefs in Electrical and Computer Engineering,
https://doi.org/10.1007/978-3-030-01024-9_3

3.2   System Model and Problem Formulation

3.2.1   System Model

We consider an MCS system shown in Fig. 3.1. The SP aims to collect the information in an area through the involvement of SUs. The SP posts the tasks through mobile phone apps. SUs who have downloaded and installed these apps can participate in the sensing procedure. Being at different locations in this area, SUs can execute different sensing tasks, and submit their sensing data to SP.

The whole sensing area is divided into m grids denoted by the set $\mathcal{G} = \{1, 2, \ldots, m\}$ and one user's sensing area can cover multiple grids, as the red dashed circle illustrated in Fig. 3.1. Considering that the significance of different locations may be different for SP, we let v_g represent the significance of the grid g, which denotes the valuation that SP can obtain after receiving the sensing data that covers the grid g. Let $\mathcal{N} = \{1, 2, \ldots, n\}$ be the set of SUs who are interested in the sensing tasks. Any user $i \in \mathcal{N}$ is associated with several attributes:

- The set $\mathcal{G}_i \subseteq \mathcal{G}$: which represents the set of grids covered by user i.
- The true cost c_i: which reflects the real cost for user i to conduct the sensing task and report the sensing data. Note that true cost c_i is the private information for user i and not open to any other users and SP.
- The bid price b_i: which is the reward that user i hopes to receive from SP to compensate the cost of conducting sensing task such as energy consumption, transmission cost, etc. Note that the bid price b_i may not be equal to the true cost c_i. For instance, user i may request a bid price higher than its true cost if it can make her achieve a higher reward.
- The reputation r_i: which reflects the quality of sensing data provided by user i. Base on the history of the received data, SP evaluates the reputation of each user.

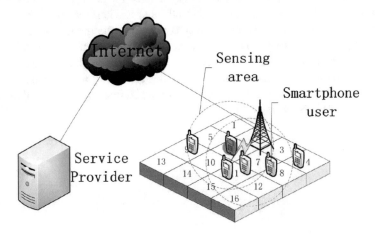

Fig. 3.1 An MCS system

Note that SP evaluates, updates, and keeps the information about the reputation of each user.

We take the quality of sensing data (i.e., the reputation of smartphone users) into consideration and use the auction theory to design an incentive mechanism. SUs interested in conducting the sensing task submit their own bid prices to SP. Based on the collected bid price, the reputation of each user, and the values of grids that each user can cover, SP determines the set of SUs to conduct the sensing task, and calculate the payment for each SU. Then, winning users execute the sensing task, submit their sensing data to SP, and receive the corresponding payment from SP.

3.2.2 Problem Formulation

Usually, the primary objective of a non-commercial SP (e.g., governmental depart-ment, a non-profit organization) is to maximize the social welfare of the system. Due to the consideration of the quality of the sensing data, we introduce the weighted social welfare, where the achieved value of the grids covered by the sensing data is weighted by the quality of the sensing data (i.e., the reputation). The weighted social welfare is defined as follows.

Definition 3.1 (Weighted Social Welfare) The weighted social welfare of an MCS system is defined as the total weighted valuation of grids covered by the sensing data that SP collects minus the total cost of SUs conducting the sensing task, which is given as.

$$S = \sum_{g \in \mathcal{G}} (v_g \times r_g) - \sum_{i \in \mathcal{N}} (b_i \times a_i) \qquad (3.1)$$

where $a_i = 1$ denotes user i is selected to perform the sensing task, otherwise $a_i = 0$. $r_g = \max\{r_h : h \in \Gamma_g\}$ represents the highest quality (i.e., reputation) among all collected data that covers the grid g, and $\Gamma_g = \{i : i \in \mathcal{N}, a_i = 1, d_i^g = 1\}$ denotes the set of all SUs who cover the grid g and are selected to conduct the sensing task. Meanwhile, $d_i^g = 1$ denotes that the grid g is covered by the sensing area of user i, otherwise $d_i^g = 0$. The notations used in this paper are listed in Table 3.1.

Definition 3.2 (Utility of Smartphone User) The utility of SU $i \in \mathcal{N}$ is defined as

$$U_i = \begin{cases} p_i - c_i & \text{if } a_i = 1 \\ 0 & \text{otherwise} \end{cases} \qquad (3.2)$$

where p_i is the payment that user i receives from SP, and c_i is the true cost of user i.

Taking the reputation of SUs into consideration, our objective is to design a truthful incentive mechanism denoted by $\Psi = (\mathcal{N}, \mathcal{G}, \mathbf{b}, \mathbf{r}, \mathbf{v})$, where given $\mathcal{N}$, $\mathcal{G}$, $\mathbf{b}$, $\mathbf{r}$ and $\mathbf{v}$, the SP determines the set of users to perform sensing tasks with

Table 3.1 Frequently used notations

Notation	Explanation
$\mathcal{N}$	The set of SUs
$\mathcal{W}$	The set of the selected SUs
$\mathcal{W}_{-i}$	The set of the selected SUs except i
$\mathcal{G}$	The set of grids
Π	The feasible set of $\mathcal{W}$
S	Weighted social welfare
$\mathcal{G}_i$	The set of grids covered by SU i
S_{-i}	Weighted social welfare excluding SU i from the system
S^*	Optimal weighted social welfare
Γ_g	The set of the selected SUs who cover grid g
$\mathcal{NC}$	The non conflict user set of current winner set $\mathcal{W}$
τ_i	The contribution set of SU i
v_g	The value of grid g
b_i	The bid price of SU i
c_i	The true cost of SU i
r_i	The reputation of SU i
p_i	The payment to SU i
U_i	The utility of SU i
$\mathbf{b}$	The bid vector
$\mathbf{r}$	The reputation vector
$\mathbf{v}$	The grid value vector
$\mathbf{p}$	The payment vector
n	The number of SUs
m	The number of grids

the objective of maximizing the weighted social welfare in (3.1). Meanwhile, SP calculates corresponding payment p_i for each user.

We adopt the weighted Vickrey-Clarke-Groves (VCG) method to design a reputation-aware incentive mechanism (RAIM). The winner selection rule and the payment rule of the proposed RAIM are elaborated in detail as follows.

3.3 Reputation-Aware Inventive Mechanism

3.3.1 Winner Selection Rule

Based on the collected bid prices, reputation, sensing coverage of each SU, and the significance of each grid, SP selects a set of SUs to perform the sensing tasks with the objective of maximizing the weighted social welfare, which can be formulated as

$$W^* = \arg\max_{W \in \Pi} S = \arg\max_{W \in \Pi}(\sum_{g \in \mathcal{G}} (v_g \times r_g(W)) - \sum_{i \in \mathcal{N}} (b_i \times a_i(W))) \qquad (3.3)$$

where W denotes a feasible selection result (i.e., a set of SUs selected to perform the sensing tasks). Π is the set of all feasible selection results that SP can adopt, and $r_g(W) = \max\{r_h : a_h(W) = 1, d_h^g = 1, h \in \mathcal{N}\}, a_i(W) = 1$ if $i \in W$.

3.3.2   Payment Rule

SP determines the payment p_i for each SU $i \in \mathcal{N}$. If SU i is not selected to perform the sensing tasks (i.e., $a_i = 0$), the payment $p_i = 0$; If user i is a winner (i.e., $a_i = 1$), the payment is:

$$p_i = b_i + (S^* - S_{-i})$$
$$= \sum_{g \in \mathcal{G}} (v_g \times r_g(W^*)) - \sum_{j \in \mathcal{N}, j \neq i} (b_j \times a_j(W^*)) - S_{-i} \qquad (3.4)$$

where S_{-i} denotes the optimal weighted social welfare when excluding user i from the auction. $r_g(W^*) = \max\{r_h : a_h(W^*) = 1, d_h^g = 1, h \in \mathcal{N}\}, a_j(W^*) = 1$ if $j \in W^*$.

Algorithm 1 RAIM($\mathcal{N}, \mathcal{G}, \mathbf{b}, \mathbf{r}, \mathbf{v}$)

1: //Stage 1: Winner selection
2: $W^* = \arg\max(\sum_{g \in \mathcal{G}} (v_g \times r_g(W)) - \sum_{i \in \mathcal{N}} (b_i \times a_i(W)))$
      $\scriptstyle W \in \Pi$
3: $S^* = (\sum_{g \in \mathcal{G}} (v_g \times r_g(W^*)) - \sum_{i \in \mathcal{N}} (b_i \times a_i(W^*)))$
4: //Stage 2: Payment determination
5: **for all** $i \in \mathcal{N}$ **do**
6:     $p_i = 0$
7: **end for**
8: **for all** $i \in W^*$ **do**
9:     $S_{-i} = \max S_{\mathcal{N} \setminus \{i\}}$
10:    $p_i = b_i + (S^* - S_{-i})$
11: **end for**
12: **return** $(S^*, W^*, \mathbf{p})$

3.3.3 Proof of Properties

We prove the proposed incentive mechanism RAIM is individually rational and truthful.

Theorem 3.3 (Individually Rational) *Our proposed incentive mechanism RAIM is individually rational.*

Proof For the SUs who are not selected as winners, their utility is zero. For each winning user $i \in \mathcal{W}^*$, his utility U_i satisfies the following equation.

$$
\begin{aligned}
U_i &= p_i - c_i \\
&= b_i + (S^* - S_{-i}) - c_i
\end{aligned}
\tag{3.5}
$$

With truthful bidding (i.e., $b_i = c_i$), $U_i = S^* - S_{-i} \geq 0$. So our proposed incentive mechanism RAIM is individually rational.

Theorem 3.4 (Truthfulness) *Our proposed incentive mechanism RAIM is truthful.*

Before we prove this theorem, we first prove several lemmas.

Definition 3.5 (Contribution Set of SU i) Given SU i is selected to conduct the sensing task, the contribution set of SU i, denoted as τ_i, is defined as the set of grids that SU i covers and has the highest reputation over all other selected users covering these grids. We have $\tau_i = \{g : g \in \mathcal{G}, r_i = \max\{r_h : h \in \Gamma_g\}\}$.

Let $\mathcal{W} = \{i, \mathcal{W}_{-i}\}$ be the set of all selected SUs, where $\mathcal{W}_{-i}$ denotes the set of all selected SUs except i.

Lemma 3.6 *Given that the set of selected SUs $\mathcal{W} = \{i, \mathcal{W}_{-i}\}$ maximizes the weighted social welfare defined in (3.3), then after excluding SU i and removing the grids in the contribution set τ_i, the SUs in the set $\mathcal{W}_{-i}$ maximize the weighted social welfare of the remaining system.*

Proof Given the winner set $\mathcal{W} = \{i, \mathcal{W}_{-i}\}$. Based on the definition of weighted social welfare, SU i has no contribution to the grids except τ_i, and all the other SUs have no contribution to the grids in τ_i. So, when we remove i and τ_i from the system, the remaining system is not affected. Therefore, the set of SUs in $\mathcal{W}_{-i}$ can still maximize the weighted social welfare of the remaining system.

Lemma 3.7 *If SU i wins by bidding b_i, he/she can also win by bidding $b_i' < b_i$.*

Proof Given that SU i wins the auction with the bid price b_i and the maximum weighted social welfare is S^*. When SU i bids a lower price $b_i' < b_i$, based on (3.3), the achieved weighted social welfare S' is larger than S^*. Since the selection rule is to maximize the weighted social welfare, SU i must still be a winner when bidding $b_i' < b_i$.

Let U_i and U_i' be the utility of SU i with truthful bid c_i and untruthful bid $b_i \neq c_i$ respectively. Based on the above lemmas, we prove Theorem 3.4 by showing that no user can improve his/her utility by bidding a price different from his/her true cost, i.e., $U_i \geq U_i'$ for any $b_i \neq c_i$. We verify all possible cases to prove the property of truthfulness.

Case 1 $b_i > c_i$

Based on Lemma 3.7, it is impossible that SU i loses by bidding c_i but wins by bidding b_i. So we need to consider three subcases:

Subcase 1.1 SU i wins with bidding both b_i and c_i. In this subcase, Let $\mathcal{W}^* = \{i, \mathcal{W}^*_{-i}\}$ be the winner set with truthful bid c_i. Then SU i's utility is

$$
\begin{aligned}
U_i &= p_i - c_i \\
&= c_i + (S(\mathcal{W}^*) - S_{-i}) - c_i \\
&= \sum_{g \in \mathcal{G}} (v_g \times r_g(\mathcal{W}^*)) - \sum_{j \in \mathcal{N}} (c_j \times a_j(\mathcal{W}^*)) - S_{-i}
\end{aligned}
\tag{3.6}
$$

where $r_g(\mathcal{W}^*) = \max\{r_h : a_h(\mathcal{W}^*) = 1, d_h^g = 1, h \in \mathcal{N}\}$, and $a_j(\mathcal{W}^*) = 1$ if $j \in \mathcal{W}^*$.

Let $\mathcal{W}' = \{i, \mathcal{W}'_{-i}\}$ be the winner set with untruthful bid b_i. Then we have

$$
\begin{aligned}
U_i' &= p_i' - c_i \\
&= b_i + (S(\mathcal{W}') - S_{-i}) - c_i \\
&= b_i + \sum_{g \in \mathcal{G}} (v_g \times r_g(\mathcal{W}')) - \sum_{j \in \mathcal{N}, j \neq i} (c_j \times a_j(\mathcal{W}')) - b_i - S_{-i} - c_i \\
&= \sum_{g \in \mathcal{G}} (v_g \times r_g(\mathcal{W}')) - \sum_{j \in \mathcal{N}, j \neq i} (c_j \times a_j(\mathcal{W}')) - c_i + c_i - S_{-i} - c_i \\
&= \sum_{g \in \mathcal{G}} (v_g \times r_g(\mathcal{W}')) - \sum_{j \in \mathcal{N}} (c_j \times a_j(\mathcal{W}')) - S_{-i}
\end{aligned}
\tag{3.7}
$$

Based on Lemma 3.6, if the winner set is $\mathcal{W}^* = \{i, \mathcal{W}^*_{-i}\}$, then the subset $\mathcal{W}^*_{-i}$ maximizes the weighted social welfare of the remaining system, which actually the system excluding SU i and removing the grids in τ_i. We assume the whole system is divided into two parts: τ_i and $\mathcal{G}_{-i}$, where $\mathcal{G}_{-i} = \mathcal{G} \setminus \tau_i$. For the part τ_i, i can maximize it. For the part $\mathcal{G}_{-i}$, $\mathcal{W}'_{-i}$ is also the winner set. So the winner set $\mathcal{W}^*$ and $\mathcal{W}'$ are same. Hence based on (3.6) and (3.7), we have $U_i' = U_i$.

Subcase 1.2 SU i wins with truthful bid c_i but loses with untruthful bid b_i. In this subcase, it is clear that $U_i' = 0$, and $U_i = p_i - c_i = c_i + (S^* - S_{-i}) - c_i = S^* - S_{-i}$. Since $S^* \geq S_{-i}$, we have $U_i \geq 0$ and $U_i \geq U_i'$.

Subcase 1.3 SU i loses with bidding both b_i and c_i. In this subcase, it is clear that $U_i' = U_i = 0$.

Case 2 $b_i < c_i$

Based on Lemma 3.7, it is impossible that SU i wins with truthful bid c_i but loses with untruthful bid b_i. So there are three subcases.

Subcase 2.1 SU i wins with bidding both b_i and c_i. For this subcase, we can follow the same analysis as the subcase 1.1 to prove $U_i' = U_i$.

Subcase 2.2 SU i loses with truthful bid c_i but wins with untruthful bid b_i. In this subcase, $U_i = 0$. Assume the optimal weighted social welfare is $S(\mathcal{W}^*)$ and $S(\mathcal{W}')$ when SU i bids c_i and b_i, respectively. Based on (3.4), when SU i bids b_i and wins the auction, his/her payment $p_i = b_i + (S(\mathcal{W}') - S_{-i})$. As we know S_{-i} is the optimal social welfare when SU i is excluded in the auction. Actually, its result is equivalent to the case when SU i loses. So $S_{-i} = S(\mathcal{W}^*)$. We can calculate the weighted social welfare (not optimal) when SU i bids c_i with the strategy $\mathcal{W}'$, and we denote this value as s. Based on the fact that $\mathcal{W}'$ is not the optimal winner set with bidding c_i but it is the optimal winner set with bidding b_i, we have $S(\mathcal{W}') = s + (c_i - b_i)$. Hence,

$$\begin{aligned}
U_i' &= p_i - c_i \\
&= b_i + (S(\mathcal{W}') - S_{-i}) - c_i \\
&= b_i + (s + (c_i - b_i) - S(\mathcal{W}^*)) - c_i \\
&= s - S(\mathcal{W}^*)
\end{aligned} \tag{3.8}$$

Because $S(\mathcal{W}^*) \geq s$, we get $U_i' \leq 0$ and $U_i' \leq U_i$.

Subcase 2.3 SU i loses with bidding both b_i and c_i. In this subcase, $U_i' = U_i = 0$.

So, based on all the cases discussed above, we can get $U_i \geq U_i'$. Hence our proposed incentive mechanism RAIM is truthful.

3.3.4 A Heuristic Algorithm

Although the proposed mechanism RAIM can achieve maximal weighted social welfare with good properties such as the truthfulness and individual rationality, the high complexity of Algorithm 1 to calculate the optimal weighted social welfare makes it different to be implemented in a large scale system. To reduce the computational complexity of the proposed mechanism RAIM, another heuristic greedy algorithm named RAIM-H is proposed. The objective of RAIM-H is to maximize the weighted social welfare with a feasible computational complexity.

The pseudo code of RAIM-H is given in Algorithm 2. It includes two stages: winner selection stage (lines 1–19) and the payment determination stage (lines 20–

Algorithm 2 $RAIM - H(\mathcal{N}, \mathcal{G}, \mathbf{b}, \mathbf{r}, \mathbf{v})$

1: //Stage 1: Winner selection
2: $\mathcal{W} \leftarrow \emptyset$;
3: $i \leftarrow \arg\max_{j \in \mathcal{N}}(r_j \cdot \sum_{k \in \mathcal{G}_j} v_k - b_j)$;
4: $a = S(\mathcal{W} \cup \{i\}) - S(\mathcal{W})$;
5: **while** $a \geq 0$ **do**
6:     $\mathcal{W} \leftarrow \mathcal{W} \cup \{i\}$;
7:     $\mathcal{NC} = \phi(\mathcal{W})$;
8:     **for all** $l \in \mathcal{NC}$ **do**
9:         $A(l) = S(\mathcal{W} \cup \{l\}) - S(\mathcal{W})$;
10:    **end for**
11:    $i \leftarrow \arg\max_{k \in \mathcal{NC}} A(k)$;
12:    **if** $\max_{k \in \mathcal{NC}} A(k) < 0$ **then**
13:        **for all** $q \in \mathcal{N} \backslash (\mathcal{NC} \cup \mathcal{W})$ **do**
14:            $A(q) = S(\mathcal{W} \cup \{q\}) - S(\mathcal{W})$;
15:        **end for**
16:        $i \leftarrow \arg\max_{q \in \mathcal{N} \backslash (\mathcal{NC} \cup \mathcal{W})} A(q)$;
17:    **end if**
18:    $a = S(\mathcal{W} \cup \{i\}) - S(\mathcal{W})$;
19: **end while**
20: //Stage 2: Payment determination
21: **for all** $i \in \mathcal{W}$ **do**
22:    $S_{-i} = \max S_{\mathcal{N} \backslash \{i\}}$
23:    $p_i = b_i + (S^* - S_{-i})$
24: **end for**
25: **return** $(\mathcal{W}, S(\mathcal{W}), \mathbf{p})$

---

24). In the winner selection stage, we select the SU with the largest weighted grid value minus bid as the first winner (line 3). Then, in order to select more SUs into the winner set, we divide all the users except $\mathcal{W}$ into two sets. The first set $\mathcal{NC}$ includes the SUs whose sensing area does not conflict with the users' in $\mathcal{W}$ (line 7). The operation $\phi(\mathcal{W})$ in line 7 is to get the non conflict set of the current winner set $\mathcal{W}$. Another set $\mathcal{N} \backslash (\mathcal{NC} \cup \mathcal{W})$ includes the SUs whose sensing area intersects with the sensing area of some SU in $\mathcal{W}$. We first search the SUs in $\mathcal{NC}$ to find a user who can induce the largest non-negative gain of the weighted social welfare, and we add this user into the winner set $\mathcal{W}$ (lines 8–11). If no such user is found, we then search the SUs in $\mathcal{N} \backslash (\mathcal{NC} \cup \mathcal{W})$ to find a user who can induce the largest non-negative gain of the weighted social welfare (lines 12–16). This search process will stop until no positive gain can obtain.

In payment determination stage, we adopt the similar payment strategy with RAIM. For each winner $i \in \mathcal{W}$, the payment p_i is calculated by applying the VCG payment rule to the proposed heuristic algorithm RAIM-H. That is, $p_i = b_i + (S^* - S_{-i})$, where S^* and S_{-i} is the weighted social warfare based on the outcome of the proposed heuristic algorithm with and without the user i in the auction, respectively.

3.4   Performance Evaluation

Simulations have been conducted to evaluate the performance of the proposed incentive mechanisms (i.e., RAIM and RAIM-H) in terms of the achieved weighted social welfare and the average reputation of SUs selected to conduct the sensing tasks. We compare RAIM and RAIM-H with three other mechanisms: random selection, reverse auction, and TSCM proposed in [3]. For the random selection, we randomly select SUs as winners and the total number of winning users is the same as that in RAIM or RAIM-H. For the reverse auction, we first sort all users' bids in ascending order, and then select the users with small bids as winners, and the number of winners is same as that in RAIM or RAIM-H as well. For TSCM, the winner selection rule is based on SUs' reputable marginal value minus their modified bids. The modified bid of an SU is his actual bid divided by his reputation value.

3.4.1   Simulation Setup

We consider an MCS system where the whole sensing area is $200\,\text{m} \times 200\,\text{m}$. This square is divided into multiple grids with $10\,\text{m} \times 10\,\text{m}$. SUs are randomly located in this area, and each user's sensing coverage includes all the grids within a distance of $20\,\text{m}$ from this user, as shown in Fig. 3.2. The bid of each user is uniformly distributed over $[1,10]$. The reputation of each user is uniformly distributed over $[0,1]$. We consider two scenarios. In the first scenario, the value of each grid is uniformly distributed over $[1,5]$. In the second scenario, there are three hotspots, in which the grids close to the center of these hotspots are much more important to the

Fig. 3.2 Simulation setup. Each red circle denotes an SU's sensing area

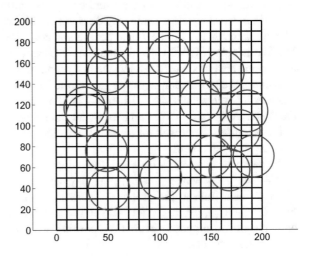

Fig. 3.3 Scenario (1) no
hotspot; Scenario (2): three
hotspots

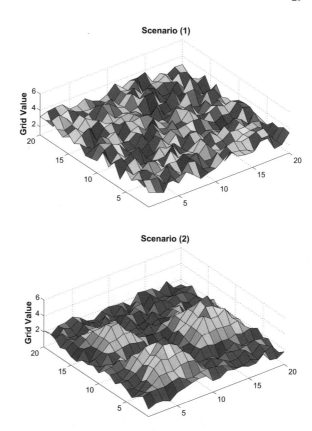

SP, so they have a larger value, as shown in Fig. 3.3. We set the average value of
grids in these two scenarios be the same.

3.4.2 Truthfulness

To verify the truthfulness of the proposed incentive mechanism RAIM, we randomly
pick up one SU (ID = 1) whose true cost is 8 and observe how his utility changes
with different bid prices from 1 to 20. Figure 3.4 shows the property of truthfulness
since the SU can not improve his utility by bidding untruthfully.

3.4.3 Weighted Social Welfare

Figure 3.5 demonstrates the impact of the number of grids on weighted social
welfare with the number of SUs $n = 15$. Compared with the random selection

Fig. 3.4 $c_1 = 8$

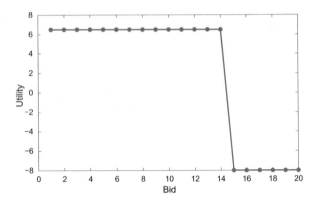

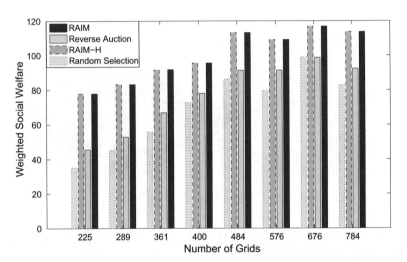

Fig. 3.5 Weighted social welfare with different numbers of grids in scenario 1, $n = 15$

and the reverse auction, the proposed mechanisms RAIM can achieve the largest weighted social welfare. Since the trends for both scenarios are the same, we only present the simulation results in scenario 1 here.

Figure 3.6 demonstrates the impact of the number of grids on weighted social welfare with more SUs ($n = 100$). It is obvious that the proposed RAIM-H outperforms other counterparts.

Figure 3.7 shows the impact of small number of users on weighted social welfare. It is obvious that the proposed RAIM and RAIM-H outperforms other counterparts. And the heuristic algorithm RAIM-H can approach to the optimal weighted social welfare obtained by RAIM very well when the number of SUs is small. Compared with TSCM and random selection, RAIM can improve the weighted social welfare by 8.65% and 48.16%, respectively, with the number of SUs $n = 16$.

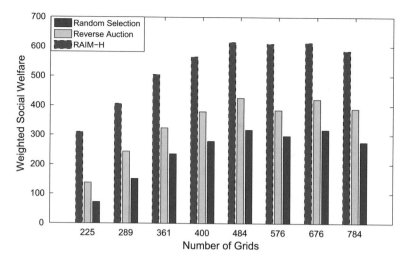

Fig. 3.6 Impact of the number of grids on weighted social welfare, $n = 100$

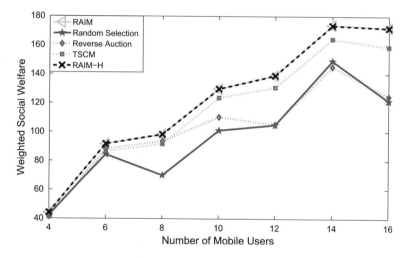

Fig. 3.7 Impact of small number of users on weighted social welfare, $m = 400$

When the number of SUs is larger, RAIM is not feasible because of the high computational complexity. So RAIM-H can work very well. Figure 3.8 shows the weighted social welfare with a larger number of SUs. Compared with the other three counterparts, RAIM-H can get the largest weighted social welfare. RAIM-H can improve the weighted social welfare by 6.15% and 75% compared with TSCM and random selection respectively, with the number of SUs $n = 100$.

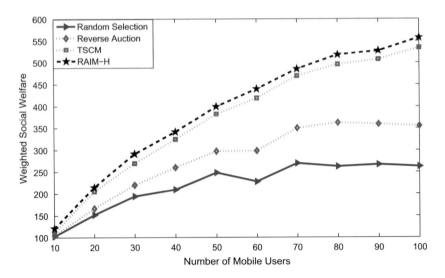

Fig. 3.8 Impact of large number of users on weighted social welfare, $m = 400$

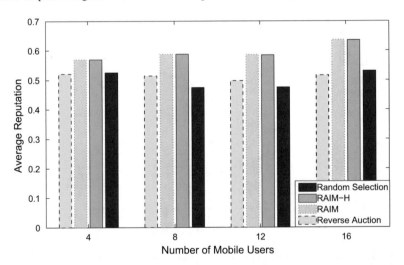

Fig. 3.9 Average reputation with different number of smartphone users, $m = 400$

3.4.4 Average Reputation

We use the average reputation over the SUs selected to conduct the sensing tasks to reflect the quality of the collected sensing data. With a larger average reputation, the system can achieve a higher quality sensing data. Figure 3.9 shows the average reputation of different mechanisms. It can be observed that the proposed mechanisms can achieve higher average reputation, which means higher quality of the collected data.

References

1. Huang KL, Kanhere SS, Hu W (2010) Are you contributing trustworthy data?: the case for a reputation system in participatory sensing. In: Proceedings of the 13th ACM international conference on Modeling, analysis, and simulation of wireless and mobile systems, 2010. ACM, pp 14–22
2. Kantarci B, Mouftah HT (2014) Reputation-based sensing-as-a-service for crowd management over the cloud. In: Communications (ICC), 2014 IEEE International Conference on, 2014. IEEE, pp 3614–3619
3. Krishna V (2009) Auction Theory, Academic press.

Chapter 4
Social-Aware Incentive Mechanism Design

4.1 An Important Factor: Social Relationship

In the previous chapter, we propose two reputation-aware incentive mechanisms RAIM and RAIM-H by taking the reputation of SUs into consideration. RAIM can guarantee the properties of truthfulness and individual rationality while maximizing the weighted social welfare. Particularly, by considering SUs' reputation, the quality of sensing data can be improved.

In addition to the quality of sensing data, the social relationship among SUs is also an important factor in the mechanism design in MCS. With the popularity of online social platforms such as WeChat, Facebook, and Twitter, SUs can easily communicate and share the information with their friends online. Therefore, the consideration of social relationship among SUs plays an important role in improving the participation level and SUs' happiness in MCS. It can be utilized to improve the performance of an MCS system. For example, allocating sensing tasks to SUs who are friends can boost the participation level in an MCS system and SUs can achieve more happiness since they can share their sensing activities and information via online social networks. In this chapter, we consider the social relationship of SUs and design a social-aware incentive mechanism to achieve higher system performance in terms of social utility and social effect. Meanwhile, the proposed incentive mechanism can achieve individual rationality, budget balance, the strongly truthfulness of SPs, and partially truthfulness of SUs.

The remainder of this chapter is organized as follows. In Sect. 4.2 we describe the system model and the problem formulation. Our proposed incentive mechanism SAIM is presented in Sect. 4.3, followed by the performance evaluation in Sect. 4.4.

F. Hou et al., *Mobile Crowd Sensing: Incentive Mechanism Design*,
SpringerBriefs in Electrical and Computer Engineering,
https://doi.org/10.1007/978-3-030-01024-9_4

4.2   System Model and Problem Formulation

4.2.1   System Model

As shown in Fig. 4.1, we consider an MCS system consisting of several SPs, multiple SUs, and one platform. These n SPs $\mathcal{N} = \{1, 2, \ldots, n\}$, and m SUs $\mathcal{M} = \{1, 2, \ldots, m\}$ are all connected to the platform via Internet, cellular network, or WiFi access.

Each SP can post its task on the platform via apps, and each SU can claim to conduct these sensing tasks via apps. Conducting the sensing task and reporting the sensing data usually consume the resource of SUs. Since SUs are rational and selfish, they will be paid as the compensation for their consumption of conducting the sensing tasks, while the SPs will issue some payment in return for the collected sensing data. Clearly, the trading between SPs and SUs should benefit both of them. That is, each SU cannot be paid less than the cost to conduct the sensing task while the payment of each SP cannot be larger than the valuation achieved from the collected sensing data. Furthermore, we will consider the social relationship of SUs and maximize the social utility of the system. Therefore, an incentive mechanism should be properly designed to allocate the sensing tasks of SPs to SUs and decide the corresponding payment.

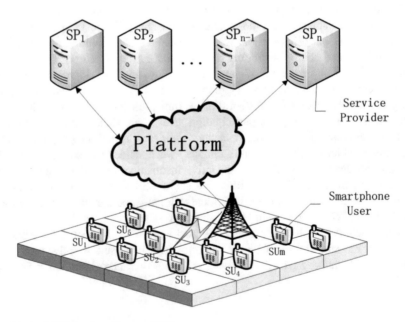

Fig. 4.1 An MCS system with several SPs

4.2.2 *Auction Problem Formulation*

We model this problem as a double auction problem, where SUs are sellers who conduct the sensing tasks and provide the sensing data, SPs are buyers who need sensing data, and the platform acts as the auctioneer who determines the allocation of sensing tasks among the SUs, charges the SPs (i.e., buyer), and pays to the SUs (i.e., seller). We assume each task can be allocated to only one SU, and each SU can conduct only one task each time. Meanwhile, all the tasks are homogeneous for SUs, which means each SU can conduct any task at the same cost.

Each SU $j \in \mathcal{M}$, as a seller who wants to conduct the sensing tasks and sells the sensing data, is associated with the following attributes:

- The true cost c_j: It denotes the real consumption of user j to conduct a sensing task and report the sensing data. This is the private information, and is not open to any SP, platform, and other SUs.
- The ask price a_j: It reflects the payment that SU j wants to receive after conducting a sensing task. Note that a_j may not be equal to c_j since an SU may ask a larger a_j to achieve a higher payment. With the truthful bidding, we have $a_j = c_j$.

Each SP $i \in \mathcal{N}$, as a buyer who wants to obtain the sensing data, is associated with the following attributes:

- The true value v_i: It reflects the real valuation that SP i can achieve after obtaining the corresponding sensing data. It is a private information, and is not open to any SU, platform, and other SPs.
- The bid price b_i: It denotes the price that SP i wants to pay after obtaining the sensing data.

In addition, we use the matrix R to reflect the social relationship of SUs, where the entity r_{ij} denotes the degree of the friendship between SU i and j. If they are not friends, the entity $r_{ij} = 0$; otherwise, $r_{ij} \in (0, 1]$. The relationship matrix R can be updated based on SUs' interaction or communications on the platform or other social networks. The discussion and the mechanism proposed in this chapter can be easily extended to other models of the social relationship.

To improve the social utility, we take the social relationship of SUs into consideration and use double auction theory to design a social-aware incentive mechanism. The process of double auction works as follows: each SP i submits the bid price b_i to the platform, and each SU j submits the ask price a_j to the platform. Meanwhile, the platform maintains the relationship matrix R by getting the relationship of all SUs. After collecting the bid price vector $\mathcal{B} = \{b_1, b_2, \ldots, b_n\}$, ask price vector $\mathcal{A} = \{a_1, a_2, \ldots, a_m\}$, and the relationship matrix R, the platform executes the task allocation and payment determination. Then the winner set of SPs and the winner set of SUs are selected, and the platform allocates the winning SPs' tasks to the corresponding SUs. After each winning SP receives the sensing data from the corresponding winning SU, it sends payment to the platform and the platform

Table 4.1 Frequently used
notations

Notation	Explanation
$\mathcal{N}$	The set of SPs
$\mathcal{M}$	The set of SUs
b_i	The bid price of SP i
v_i	True value for the task of SP i
n	The number of SPs
m	The number of SUs
a_j	The ask price of SU j
c_j	The cost of SU j
r_{ij}	The closeness between SUs i and j
R	The relationship matrix of all SUs
$\mathcal{W}^b$	The winner set of SPs
$\mathcal{W}^s$	The winner set of SUs
$\mathcal{B}$	The bid price vector of SPs
$\mathcal{A}$	The ask price vector of SUs
U_s	Social utility
U_j^s	The utility of SU j
U_i^b	The utility of SP i
S_w	Social welfare
S_m	Social effect
p_j^s	The payment of SU j
p_i^b	The charge of SP i
λ	Social effect parameter

sends the corresponding payment to the winning SUs. This completes the whole participatory sensing process. The proposed incentive mechanism including the winner selection rule and payment rule will be elaborated as follows. The main notations used in this chapter are given in Table 4.1.

4.3 Social-Aware Incentive Mechanism

The objective of the proposed social-aware incentive mechanism SAIM is to maximize the social utility by selecting appropriate winning SPs and winning SUs while designing the proper payment rule to ensure the system efficiency and nice properties such as truthfulness, individual rationality, and budget balance. Before presenting the proposed mechanism SAIM in detail, we introduce some definitions first.

Definition 4.1 (Social Welfare) The social welfare of an MCS system is defined as the total valuation that SPs can get after obtaining the sensing data minus the total cost of SUs who conduct the sensing tasks, which is given as:

$$S_w = \sum_{i \in \mathcal{W}^b} b_i - \sum_{j \in \mathcal{W}^s} a_j, \tag{4.1}$$

where $\mathcal{W}^b$ is the winner set of SPs (i.e., buyers), and $\mathcal{W}^s$ is the winner set of SUs (i.e., sellers).

Definition 4.2 (Social Effect) The social effect is defined to be proportional to the sum of all SUs' relationship, which is given as:

$$S_m = \frac{\lambda}{2} \sum_{u,v \in \mathcal{W}^s} r_{uv}, \tag{4.2}$$

where λ is a parameter associated with social effect, and r_{uv} denotes the social relationship between SUs u and v.

Definition 4.3 (Social Utility) The social utility is defined as the sum of social welfare and social effect, which is given as:

$$U_s = \sum_{i \in \mathcal{W}^b} b_i - \sum_{j \in \mathcal{W}^s} a_j + \frac{\lambda}{2} \sum_{u,v \in \mathcal{W}^s} r_{uv} \tag{4.3}$$

Definition 4.4 (Utility of SU) The utility of SU $j \in \mathcal{M}$ is defined as:

$$U_j^s = \begin{cases} p_j^s - c_j & \text{if } j \in \mathcal{W}^s \\ 0 & \text{otherwise} \end{cases} \tag{4.4}$$

where p_j^s is the achieved payment of SU j, and c_j is the true cost of SU j.

Definition 4.5 (Utility of SP) The utility of SP $i \in \mathcal{N}$ is defined as:

$$U_i^b = \begin{cases} v_i - p_i^b & \text{if } i \in \mathcal{W}^b \\ 0 & \text{otherwise} \end{cases} \tag{4.5}$$

where v_i is the true value for the task of SP i, and p_i^b is the price SP i needs to pay.

4.3.1 Incentive Mechanism SAIM

The proposed social aware incentive mechanism SAIM includes two parts: the first part is the winner selection rule which determines how to select the set of winning SUs to conduct the sensing tasks and the set of SPs which can successfully collect the sensing data; the second part is the payment rule which determines the payment to SUs and the charge from SPs.

4.3.2 Winner Selection Rule

We sort the bid prices of all SPs in the decreasing order and sort the ask prices of all SUs in the increasing order as follows:

$$b_1 \geq b_2 \geq \cdots \geq b_n \tag{4.6}$$

$$a_1 \leq a_2 \leq \cdots \leq a_m \tag{4.7}$$

After sorting, we find the largest k such that $b_k \geq a_k$ and $b_{k+1} \leq a_{k+1}$. Then, we fix b_k and find the largest h such that $b_k \geq a_h$ and $b_k \leq a_{h+1}$. To ensure the system efficiency, we choose the first $(k - 1)$ SPs as the winning SPs (i.e., buyers) and include them into the winner set $\mathcal{W}^b$. Then we select $(k - 1)$ SUs from $(h - 1)$ as the winning SUs with objective to maximize the social utility. The maximization problem can be formulated as follows:

$$\max_{\mathcal{W}^s \in \Pi} U_s = \max_{\mathcal{W}^s \in \Pi} \left(\sum_{i \in \mathcal{W}^b} b_i - \sum_{j \in \mathcal{W}^s} a_j + \frac{\lambda}{2} \sum_{u,v \in \mathcal{W}^s} r_{uv} \right) \tag{4.8}$$

where Π denotes the set of all feasible winner sets for SUs and $\mathcal{W}^b$ is the set of winning SPs.

After $\mathcal{W}^b$ is decided based on the aforementioned method, the problem (4.8) can be rewritten as a binary quadratic programming (BQP) problem as follows:

$$\max_{\mathcal{W}^s \in \Pi} \left(\tfrac{\lambda}{2} x^T R x - A^T x \right) \tag{4.9}$$

s.t.

$$e^T x = t$$

$$x_i \in \{0, 1\}, \quad i = 1, 2, \ldots, (h - 1)$$

where $x = (x_1, x_2, \ldots, x_{h-1})^T$, and $x_i = 1$ means SP i is selected as a winner, otherwise $x_i = 0$. $A = (a_1, a_2, \ldots, a_{h-1})^T$ is the ask vector, $e = (1, 1, \ldots, 1)^T$ is a vector with all $(h - 1)$ elements 1. R is the relationship matrix, t is the number of winning SPs (i.e., $t = |\mathcal{W}^b|$).

When we set the vector $A = 0$, the problem (4.9) is reduced to the form of Maximum Cut problem which has been proved to be NP-hard [1]. Therefore, the problem (4.9) is a NP-hard problem, which means that the optimal winner set and the maximum social utility can not be solved in polynomial time.

We design a heuristic algorithm to select the winner set $\mathcal{W}^s$, which is shown in Algorithm 1. The input parameters include $\mathcal{W}^{s'}$ (i.e., the set of the first $(h - 1)$ SUs) and $\mathcal{W}^b$ (i.e., the set of the selected winning SPs). Starting from the SU with the minimum ask price in $\mathcal{W}^{s'}$, we get the friend set of this SU (line 5). The operation $\mathcal{F} = \phi(\mathcal{W}^s)$ is to get the friends of all SUs in $\mathcal{W}^s$. Then we select an SU in $\mathcal{F}$

which can induce the maximum gain of the social utility into the winner set $\mathcal{W}^s$. We calculate the social utility gain due to adding SU i (line 14). $f(\mathcal{W}^{s''})$ denotes the social utility of set $\mathcal{W}^{s''}$. Then we get a new winner set $\mathcal{W}^s$, and we repeat this procedure until the number of winners in $\mathcal{W}^s$ is equal to the number of winners in $\mathcal{W}^b$. The computational complexity of Algorithm 1 is $O(n^2)$.

Algorithm 1 HeuristicGreedy($\mathcal{W}^{s'}, \mathcal{W}^b$)

1: //Stage 1: Initialization
2: $\mathcal{W}^s \leftarrow \emptyset, \mathcal{F} \leftarrow \emptyset$;
3: $i \leftarrow \arg\min_{j \in \mathcal{W}^{s'}} a_j$;
4: $\mathcal{W}^s \leftarrow \mathcal{W}^s \bigcup\{i\}$;
5: $\mathcal{F} \leftarrow \phi(\mathcal{W}^s)$;
6: **while** ($|\mathcal{W}^s| \neq |\mathcal{W}^b|$) **do**
7:     $\mathcal{F} \leftarrow \phi(\mathcal{W}^s)$;
8:     **if** $\mathcal{F} = \emptyset$ or $\mathcal{F} \cap \mathcal{W}^s = \emptyset$ **then**
9:         $l = \arg\max_{j \in \mathcal{W}^{s'} \setminus \mathcal{W}^s} a_j$;
10:        $\mathcal{F} \leftarrow \mathcal{F} \cup \{l\}$
11:     **end if**
12:     **for all** $i \in \mathcal{F}$ **do**
13:         $\mathcal{W}^{s''} \leftarrow \mathcal{W}^s \cup \{i\}$
14:         $G(i) = f(\mathcal{W}^{s''}) - f(\mathcal{W}^s)$
15:     **end for**
16:     $g = \arg\max_{i \in \mathcal{F}} G(i)$;
17:     $\mathcal{W}^s \leftarrow \mathcal{W}^s \cup \{g\}$;
18: **end while**
19: **return** ($\mathcal{W}^s$)

4.3.3 Payment Rule

After deciding $\mathcal{W}^b$ and $\mathcal{W}^s$, the platform needs to determine the payment to each SU (i.e., p_j^s) and the charge to each SP (i.e., p_i^b). To guarantee the nice properties such as truthfulness, individual rationality, and budget balance, the platform charges all winning SPs b_k and pays all winning SUs a_h.

In summary, the pseudo code of the proposed incentive mechanism SAIM is described in Algorithm 2: lines 1–5 are initialization; Lines 6–8 are to determine the winner sets of SPs while lines 9–10 is to determine the winner set of SUs; and lines 12–13 are to determine the payment to SUs and the charge to SPs. The proposed social-aware incentive mechanism SAIM takes the social relationship into account. With the consideration of the social relationship of SUs, the proposed mechanism more likely allocates the sensing tasks to SUs with friendship, which can achieve a higher social utility. Meanwhile, SAIM can guarantee truthfulness, individual rationality, and budget balance.

Algorithm 2 SAIM($\mathcal{N}, \mathcal{M}, \mathcal{B}, \mathcal{A}$)

1: $\mathcal{W}^b \leftarrow \emptyset, \mathcal{W}^s \leftarrow \emptyset$
2: Sort bid prices of SPs in descending order: $b_1 \geq b_2 \geq, \cdots \geq b_n$;
3: Sort all prices of SUs in increasing order: $a_1 \leq a_2 \leq, \cdots \leq a_m$;
4: Find the largest k, so that $b_k \geq a_k$ and $b_{k+1} \leq a_{k+1}$;
5: Find the largest h, so that $b_k \geq a_h$ and $b_k \geq a_{h+1}$;
6: //Determine the winning SPs
7: $\mathcal{W}^b \leftarrow$ the first $(k-1)$ SPs;
8: $\mathcal{W}^{s'} \leftarrow$ the first $(h-1)$ SUs;
9: //Determine the winning SUs using the heuristic greedy algorithm
10: $\mathcal{W}^s \leftarrow$ HeuriticGreedy($\mathcal{W}^{s'}, \mathcal{W}^b$);
11: //Determine the charge and payment
12: $p_i^b = b_k, \forall i \in \mathcal{W}^b$;
13: $p_j^s = a_h, \forall j \in \mathcal{W}^s$;
14: **return** $(\mathcal{W}^b, \mathcal{W}^s, p_i^b, p_j^s)$

4.3.4   Proof of Properties

The proposed mechanism SAIM satisfies several nice properties including individual rationality, budget balance, and truthfulness.

Theorem 4.6 (Individually Rational) *SAIM is individually rational.*

Proof The individual rationality means that each SP and SU will have a non-negative utility.

For each winning SP $i \in \mathcal{W}^b$, his utility satisfies $U_i^b = b_i - p_i^b = b_i - b_k$. Because SP i is the winner, based on the sorted sequence in (4.6), $b_i \geq b_k$, so $U_i^b \geq 0$. For each winning SU $j \in \mathcal{W}^s$, the utility satisfies $U_j^s = p_j^s - a_j = a_h - a_j \geq 0$. For the SPs and SUs who are not winner, their utility is 0. So our proposed incentive mechanism is individually rational.

Theorem 4.7 (Budget Balance) *SAIM is budget-balanced.*

Proof To prove the budget balance of an auction, we need to prove the total expense of SPs is no less than the total payment to SUs.

Because $|\mathcal{W}^s| = |\mathcal{W}^b| = t$, we have $\sum_{i \in \mathcal{W}^b} p_i^b - \sum_{j \in \mathcal{W}^s} p_j^s = t p_j^s - t p_j^s = t(b_k - a_h) \geq 0$. Since we have $b_k \geq a_h$, the total expense of SPs is no less than the total payment to SUs. This property holds.

Theorem 4.8 (Truthfulness) *SAIM is completely truthful for SPs and partially truthful for SUs.*

To prove the truthfulness of our proposed incentive mechanism SAIM, we need to prove it is truthful for SPs (i.e., buyers) and SUs (i.e., sellers), respectively. Before the proof, we give several lemmas first.

Lemma 4.9 (Monotone for SUs) *The selection rule is monotone for SUs. That is, if SU j wins the auction by asking a_j, she can also win by asking $\overline{a}_j \leq a_j$*

Proof Based on the sorted rule (i.e., lines 2–5 in Algorithm 2), when SU j asks a_j and wins the auction, it denotes that the index of a_j in the sorted sequence (4.7) is before index h. Because $\overline{a}_j \leq a_j$, the boundary value h is exactly the same for asking $\overline{a}_j$ and a_j. So the index of $\overline{a}_j$ is also before h. Based on Algorithm 1, if SU j has the minimum ask price, i.e., $j = \arg\min_{i \in \mathcal{W}^{s'}} a_i$, then j is the winner definitely. In this situation, if SU j asks $\overline{a}_j$, she also has the minimum ask price, so she also wins the auction. Otherwise, based on the selection rule lines 12–17 in Algorithm 1, SU j can induce much more social utility gain by asking $\overline{a}_j \leq a_j$, which implies SU j is also a winner.

Lemma 4.10 (Monotone for SP) *The selection rule is monotone for SPs, i.e., if SP i wins the auction by bidding b_i, she can also win by bidding $\overline{b}_i \geq b_i$.*

Proof When SP i bids b_i and wins the auction, it denotes that the index of b_i is before k in the sorted sequence (4.6). Since $\overline{b}_i \geq b_i$, the boundary value k is exactly the same for bidding $\overline{b}_i$ and b_i. Definitely the index of $\overline{b}_i$ is also before k. So SP i is also a winner.

Lemma 4.11 *If SU j wins the auction by asking a_j and $\overline{a}_j$, she will get the same payment, i.e., $p_j^s = \overline{p}_j^s$.*

Proof Based on the sorted rule lines 2–5 in Algorithm 2, when SU j wins with both a_j and $\overline{a}_j$, the boundary value h of these two situations is the same. It means that the order in the sorted sequence (4.7) after a_{h-1} is exactly the same. So, SU j will get the same payment $p_j^s = \overline{p}_j^s = a_h$.

Lemma 4.12 *If SP i wins the auction by bidding b_i and $\overline{b}_i$, then she will pay the same price, i.e., $p_i^b = \overline{p}_i^b$.*

Proof Based on the sorted rule lines 2–3 in Algorithm 2, if SP i wins the auction by bidding both b_i and $\overline{b}_i$, it means that b_i and $\overline{b}_i$ are both in the front of the boundary value b_k. Because the boundary value will not change in these two cases, SP i will pay the same price $p_i^b = \overline{p}_i^b = b_k$.

Lemma 4.13 *SAIM is completely truthful for service providers.*

To prove Lemma 4.13, we just need to show that no SP can improve her utility by bidding different from her true value, i.e., $\overline{U}_i^b \leq U_i^b$ for any $b_i \neq v_i$, where $\overline{U}_i^b$ and U_i^b are the utility of SP i with the bid price b_i and v_i respectively.

Proof We will depict two possible cases to prove the truthfulness of SPs.

Case 1 $b_i > v_i$

In this case, based on Lemma 4.10, it is impossible that SP i wins with the bid price v_i but loses with the bid price b_i. So we discuss the left three subcases.

Subcase 1.1 SP i wins the auction when bidding both b_i and v_i. Based on Lemma 4.12, when SP i wins with different bid prices, the payment is the same. So $p_i^b = \overline{p}_i^b$. Then, we have $\overline{U}_i^b = v_i - \overline{p}_i^b = v_i - p_i^b = U_i^b$.

Subcase 1.2 SP i loses when bidding both b_i and v_i. In this subcase, $\overline{U}_i^b = U_i^b = 0$.

Subcase 1.3 SP i wins with bid price b_i but loses with the bid price v_i. In this case, $\overline{U}_i^b = v_i - \overline{p}_i^b = v_i - b_k$. Because i loses with the bid price v_i, we can get $v_i \leq b_k$. So $\overline{U}_i^b \leq 0 = U_i^b$.

Case 2 $b_i < v_i$

In this case, based on Lemma 4.10, it is impossible that SP i wins with the bid price b_i but loses with the bid price v_i is impossible. So we only discuss other three subcases.

Subcase 2.1 SP i wins the auction when bidding both b_i and v_i. Based on Lemma 4.12, when SP i wins with different bid prices, the payment is the same. So $p_i^b = \overline{p}_i^b$. Therefore, we have $\overline{U}_i^b = v_i - \overline{p}_i^b = v_i - p_i^b = U_i^b$.

Subcase 2.2 SP i loses when bidding both b_i and v_i. In this subcase, $U_i^b = \overline{U}_i^b = 0$.

Subcase 2.3 SP i wins with the bid price v_i but loses with the bid price b_i. When SP i bids b_i and loses the auction, b_i must be behind of b_k, and it is obvious that $\overline{U}_i^b = 0$. So we can get $U_i^b = v_i - p_i^b = v_i - b_k \geq 0$. That is, $U_i^b \geq \overline{U}_i^b$.

Lemma 4.14 *SAIM is partially truthful for SUs.*

We will show no SU can improve her utility by bidding different from her true cost in most cases, i.e., $\overline{U}_j^s \leq U_j^s$ for any $a_j \neq c_j$. We discuss all possible cases to prove this lemma.

Proof There are also two possible cases we need to discuss.

Case 1 $a_j \geq c_j$

In this case, based on Lemma 4.9, it is impossible that SU j wins with the ask price a_j but loses with the ask price c_j. So we just need to discuss three subcases as follows:

Subcase 1.1 SU j wins the auction when asking both a_j and c_j. Based on Lemma 4.11, we have $\overline{p}_j^s = p_j^s$. So $\overline{U}_j^s = \overline{p}_j^s - c_j = p_j^s - c_j = U_j^s$.

Subcase 1.2 SU j loses when asking both a_j and c_j. In this subcase, it is obvious that $U_j^s = \overline{U}_j^s = 0$

Subcase 1.3 SU j wins with the ask price c_j but loses with the ask price a_j. Because SU j wins with c_j, the utility $U_j^s = p_j^s - c_j = a_h - c_j \geq 0$. It is clear that $\overline{U}_j^s = 0$ because j loses with the ask price a_j. So we have $U_j^s \geq \overline{U}_j^s$

Case 2 $a_j < c_j$

Based on Lemma 4.9, it is impossible SU j loses with the ask price a_j but wins with the ask price c_j. So, there are three subcases as follows:

Subcase 2.1 SU j wins when asking both a_j and c_j. Based on Lemma 4.11, we have $\overline{p}_j^s = p_j^s$. So $\overline{U}_j^s = \overline{p}_j^s - c_j = p_j^s - c_j = U_j^s$.

Subcase 2.2 SU j loses when asking both a_j and c_j. In this subcase, it is obvious that $U_j^s = \overline{U}_j^s = 0$.

Subcase 2.3 SU j wins with the ask price a_j but loses with the ask price c_j. In this subcase, if $c_j \geq a_h$ we can guarantee the truthfulness. Since $a_h \leq c_j$ in this situation, we have $\overline{U}_j^s = \overline{p}_j^s - c_j = a_h - c_j <= 0 = U_j^s$. When $a_j \leq c_j \leq a_h$, we can not guarantee truthfulness since $\overline{U}_j^s = a_h - c_j \geq 0 = U_j^s$. However, this situation rarely happens since most SUs are less likely to ask a price lower than their true cost. So, the influence of this case is very small.

Based on Lemmas 4.13 and 4.14, we can prove Theorem 4.8.

4.4 Performance Evaluation

Extensive simulations have been conducted to evaluate the performance of SAIM in terms of the social utility, the social effect, and other properties. We compare SAIM with the optimal social utility, McAfee double auction, and random selection. For random selection, we randomly select SPs and SUs as winners and the number of winners is the same as that with SAIM.

We consider the bid price of each SPs is uniformly distributed over [5, 20]. The ask price of each SUs is normally distributed with the mean of 13 and the standard deviation of 1.

4.4.1 Truthfulness

To show the truthfulness of SPs, we randomly pick up one SP (ID = 4) whose true value is 17.37. We observe how his utility changes when he bids different prices between 1 and 22. Figure 4.2 shows the property of truthfulness the SP cannot improve his utility by bidding a price different from his true value.

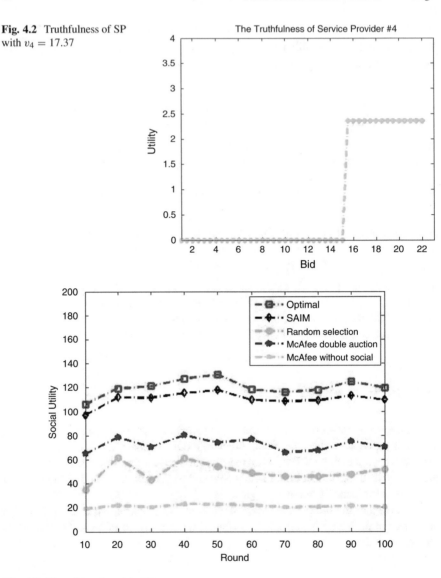

Fig. 4.2 Truthfulness of SP with $v_4 = 17.37$

Fig. 4.3 The achieved social utility

4.4.2   Social Utility

Figure 4.3 shows the achieved social utility. It is observed that SAIM can achieve larger social utility compared with McAfee double auction and random selection. Meanwhile, SAIM can approach to the optimal value very well. Figure 4.4 shows the impact of the number of SPs on the social utility with the number of SUs $m = 100$.

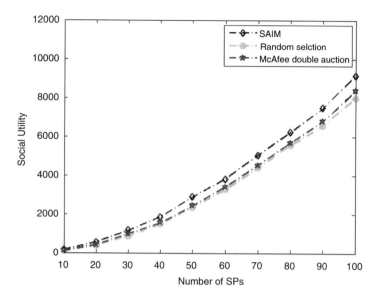

Fig. 4.4 Impact of n with $m = 100$

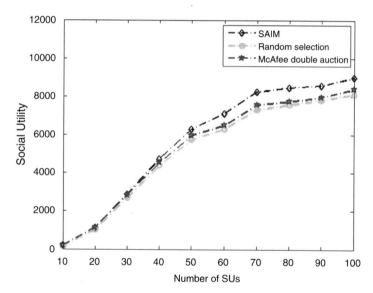

Fig. 4.5 Impact of m with $n = 100$

It is observed that the social utility increases with the increase of n. This is due to the increase of the number of winners. Figure 4.5 shows the impact of the number of SUs on social utility with the number of SPs $n = 100$.

4.4.3 Social Effect

Figures 4.6 and 4.7 show the achieved social effect with different number of SPs and the different number of SUs, respectively. It is observed that our proposed SAIM achieves larger social effects. With the consideration of social relationship, SAIM more likely allocates tasks to SUs with a close relationship, which leads to larger social effects.

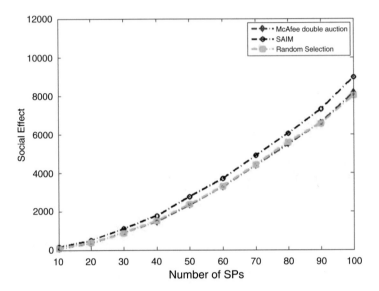

Fig. 4.6 The achieved social effect with $m = 100$

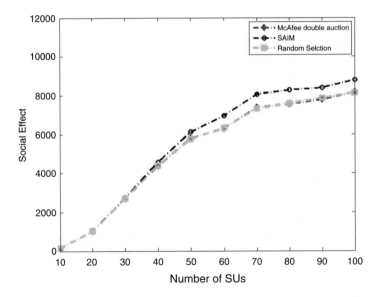

Fig. 4.7 The achieved social effect with $n = 100$

Reference

1. Garey M. R., Johnson D. S. (1979) Computers and Intractability: A Guide to the Theory of NP-Completeness, W. H. Freeman and Company, New York.

Chapter 5
Conclusions

5.1 Summary

This book firstly investigates and reviews the development and various applications of MCS. With the miniaturization of sensors and the popularity of smart mobile devices, MCS becomes a promising solution to efficiently collect different types of information such as traffic condition, air quality, temperature, etc. In addition, The auction theory and the incentive mechanism design are briefly introduced. Incentive mechanism design plays a key role in the success of MCS. An efficient incentive mechanism is a key factor to attract enough number of SUs to participate in an MCS system, then enough number of high-quality sensing data can be collected. Efficiently addressing incentive mechanism design enables the success of MCS applications such as air quality monitoring and dynamic traffic monitoring. These applications are tightly related to our daily life and society, especially the achievement of the smart city and Internet of Things.

Two types of incentive mechanisms with different system models are introduced in this book. One is the reputation-aware incentive mechanism, and another is the social-aware incentive mechanism. For the former one, we have presented the incentive mechanisms RAIM and RAIM-H, where we consider an MCS system consisting of one SP, multiple SUs, and one platform. We introduce the reputation of each SU to reflect the quality of sensing data provided by each SU. By taking the reputation into consideration, we design a reputation-aware incentive mechanism RAIM to maximize the weighted social welfare and improve the quality of the collected sensing data. Our proposed incentive mechanism RAIM is truthful and individually rational. In addition, incentive mechanism RAIM-H is designed to reduce the computational complexity. The simulation results have demonstrated the better performance of both RAIM and RAIM-H compared with other three counterparts in terms of weighted social welfare, average reputation and truthfulness. For the latter one, we have proposed incentive mechanism SAIM, where we consider an MCS system composed of multiple SPs, multiple SUs, and one platform. We take the social relationship of SUs into consideration, and design a social-aware incentive mechanism SAIM to obtain a larger social utility and higher happiness of SUs. SAIM can guarantee individual rationality, budget balance, strongly truthfulness for SPs and partially truthfulness for SUs. Meanwhile, it can approach to the maximal social utility very well. Extensive simulations have been conducted to demonstrate the good performance of SAIM and the impact of different parameters on the performance in terms of social utility and social effect.

5.2 Future Work

For MCS, there still exist several challenging issues such as cooperative sensing, group sensing, fairness in the incentive mechanism design, etc. We elaborate some of them as follows.

- *Cooperative sensing*: Cooperative sensing is an important way to benefit both SUs and SPs. In an MCS system, a sensing task may be composed of multiple requirements. Meanwhile, different SUs may be good at different requirements. For example, a whole-day noise monitoring program requires the sensing data from the morning to the midnight. Some users such as sanitation workers are available in the morning and inconvenient at night, while other users like night shifts' schedule are reverse. For every individual user, it is hard to accomplish the whole task since the cost is very high. Therefore, it is a good choice for heterogeneous users to cooperate with each other and work as a group to obtain a larger surplus.
- *Group sensing*: Group sensing is another interesting research issue in MCS. Data is more valuable nowadays than anytime before. Customers' preference plays a significant role for a large number of companies such as Alibaba and Tencent. A large volume of data helps them to develop more popular products. For a data collector, group sensing data is much more valuable than individual data. Thus the data collector is willing to provide a higher unit price for a group of sensing data than separated one. If multiple users can form a group and sell their sensing data together, they will obtain more benefits. The main challenge is how to form groups and how to share the payoff among group members fairly, efficiently, and individual rationally.
- *Diversity of mobile users*: The diversity of users is also worth to be considered into MCS applications. Users with different backgrounds (e.g., location, age, gender, education background, etc.) convey more valuable information than users with a similar background. By further considering the diversity of users, the sensing quality can be improved.

Printed in the United States
By Bookmasters